AC Power Conditioners

AC Power Conditioners

DESIGN AND APPLICATIONS

JAMES W. CLARK

ACADEMIC PRESS, INC.
Harcourt Brace Jovanovich, Publishers
San Diego New York Berkeley Boston London Sydney Tokyo Toronto

Academic Press, Inc.
San Diego, California 92101

United Kingdom Edition published by
Academic Press Limited
24–28 Oval Road, London NW1 7DX

Library of Congress Cataloging-in-Publication Data

Clark, James W.
AC power conditioners : design and applications / James W. Clark.
p. cm.
Includes index.
ISBN 0-12-175460-X (alk. paper)
1. Electric power supplies to apparatus. 2. Electric apparatus and appliances--Protection. I. Title.
TK4015.C58 1990
621.381'044--dc20 89-34783
CIP

Printed in the United States of America
89 90 91 92 9 8 7 6 5 4 3 2 1

Contents

Preface

Many good books have been written in the past decade dealing with the subject of solid state power conversion. The bulk of those books deal with the techniques involved in designing DC switching power supplies. AC power conversion utilizes many of the same techniques as DC power supply design but with the difference that an AC output results. AC power line conditioners can take on any one of, or a combination of, several forms. In the simplest form an AC line conditioner will provide EMI/RFI filtering, while the most complex form will completely replace the AC power line.

The purpose of this book is to provide the reader first with some insight into the various types of power line disturbances and problems that occur daily, then provide some examples of how to go about solving those problems. It was not written to further educate engineers already experienced in power conversion—although the possibility does exist—but more for those who are less experienced and wish to gain more knowledge of the subject.

Industry has grown to depend greatly on the presence of reliable AC power to run the equipment necessary for everyday business. Computers, copy machines, typewriters, lighting, and so on, all operate from electricity, usually provided by a power company. When the lights go out, most businesses stop until power is restored. Even emergency lighting does little good since most jobs performed by office and factory workers today require equipment that operates from the AC line. In addition to the lost productivity caused by power disturbances, there is also the risk of lost data or damage to equipment. An engineer with little or no experience in power conversion should be able to follow the explanations and examples in this book and come away with the ability to design and specify equipment as required to provide power protection for any situation. While it may not be practical to try and provide emergency power for an entire business, it may be possible to provide line conditioners or backup power to critical equipment, thus protecting vital data and protecting expensive equipment against damage. Having read this book you will be able to evaluate situations and make informed decisions about what sort of protection is required.

This book is divided into four parts, each dealing with a different form of power protection. Part I will introduce you to the various types of power problems encountered. Each part becomes progressively more complex in its discussions as the problem becomes more complex. You should begin with the first part and read the complete book rather than seeking out information only on a particular subject. This will provide a more structured introduction to the whole area of AC power conditioners and how to select the best solution to a particular problem. The last part, dealing with protection against a complete loss of AC power, or blackout, is the most complex and covers a lot of ground. Devices that provide emergency backup power are known as uninterruptible power sources. To successfully design this type of product requires knowledge of many different subjects. The basic operation of power transistors is covered since they are commonly utilized to do the actual switching in the converter. Batteries are used to store energy for the converter to draw from during power failure, so a discussion of batteries and their selection is included. Output filter design is necessary to provide the waveshape required by the load. A section on filter theory is included with examples. Cooling is important to get the best reliability, so a section on cooling techniques is also included.

After having read this book I am sure any person having an understanding of the subject material will be better prepared to deal with the problems of everyday power line disturbances.

PART I

Introduction to Power Problems

CHAPTER 1
Introduction

Ever since the early days of alternating-current (AC) voltage generation to homes and industry, the reliability and dependence on that voltage being there, and staying within some predictable limits, has increased tremendously. The power generating industry has increased the reliability of its product by constantly updating equipment and automating controls with backup systems in an effort to satisfy the ever-increasing demands placed on it. With the increased reliability of commercially generated power, people have grown to depend on it for everything from entertainment to national defense. This continued dependence, and the growth of industries that rely on electricity, have imposed such a strain on the power available that it is sometimes difficult to obtain clean, reliable AC power for routine operation.

Consider for a moment that other than AC motors there is very little in the way of useful equipment that operates directly from the 120-V-AC power line. Generally there is some sort of buffer between the power line and the actual brains of the equipment it operates, such as the power supply found in computers. This buffer is relied on to modify the raw AC voltage into something more predictable and steady in order to properly operate the equipment of which it is a part. Unfortunately, the designers of this equipment don't often know of the evils that lurk at the wall outlet. The IEEE (Institute of Electrical & Electronics Engineers) Guide for Surge Voltages in Low-Voltage AC Power Circuits (IEEE Standard 587) indicates that amplitudes in excess of 6000 V may appear at the wall outlet; the limitation is that the outlet will arc over and limit the voltage, with enough current in this voltage to do real damage to most equipment that may be connected to it.

The American National Standard (ANSI C84.1) for electric power systems and equipment shows a 120-V service as having an acceptable range of 108–126 V by the time the voltage reaches the user. This includes drops in internal wiring. The same standard also states that momentary excursions beyond these limits are not normal, but are expected. The duration and extent of these excursions are not limited by any standard. The quality of the power reaching critical equipment often has little to do with the quality of the power arriving at the service entrance to a building. In addition to disturbances that can appear on the power line ahead of the service entrance, additional problems can occur within the building that can affect the operation of critical equipment (see Fig. 1-1).

The first step to solving any power-related problem is to ensure the installation of the equipment, say, a computer, for example, is made in a manner that will minimize the effect of the operation of other equipment on the computer. Figure 1-1 illustrates an example of how the power service enters and is distributed throughout a building. An installation that could be prone to significant problems is shown in Fig. 1-2*a*. It is perfectly normal for two adjoining rooms to share a common power line from the distribution panel. The operation of a computer that is connected to a power line, which also operates equipment containing compressors or large motors, could be affected by the operation of that equipment. A more desirable installation is shown in Fig. 1-2*b* where a dedicated power line is installed from the distribution panel to the computer.

Several types of disturbance will occur on the power lines and can result in anything from improper operation to damage in equipment. Figure 1-3 shows the effects of these disturbances on the power-line waveform. These disturbances are noise or spikes, surge, sag, and complete blackout. Each of these disturbances can be the result of one or more of many different causes, both internal and external to the building. The example shown in Fig. 1-2*a* is a

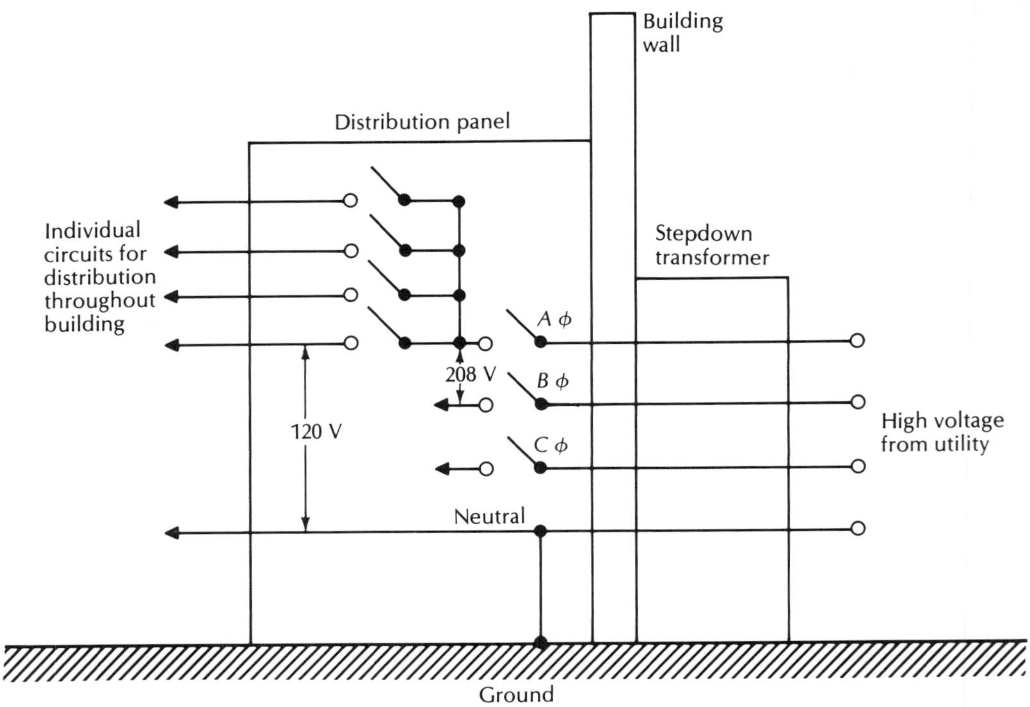

FIGURE 1-1 Typical service entrance to building.

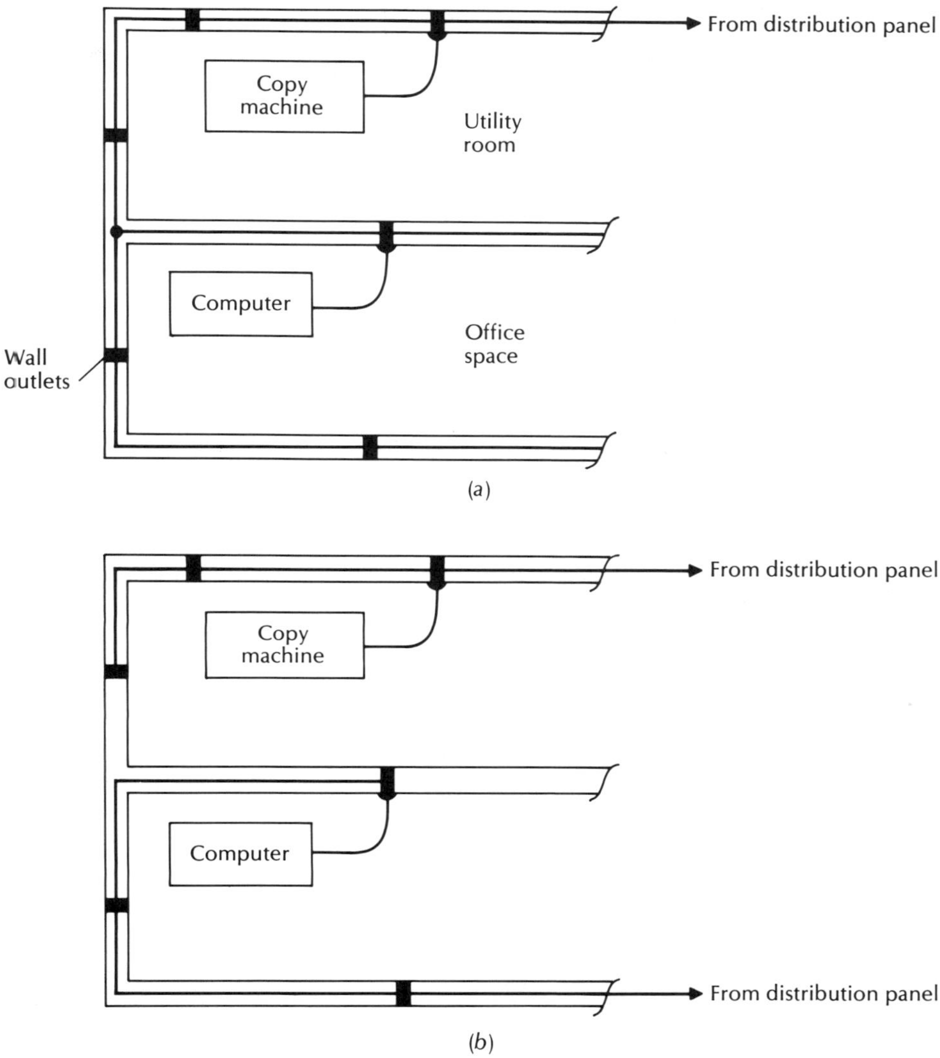

FIGURE 1-2 Computer (*a*) shares common power line with copy machine; (*b*) computer is powered from dedicated line, removing possibility of the copy machine affecting operation of computer.

case where a sag can be caused internal to the building, as a result of a compressor starting, while acceptable power is being delivered from outside the building. Another cause of both sag and surge voltage is ground faults. Ground faults are caused whenever a path is provided from a high-voltage line to ground. Typically when a ground fault occurs there will be an abrupt change in voltage, not only on the faulted line, but also on the other phase lines and ground. This abrupt change in voltage, even when occurring several miles from the transformer at the building, will cause what is known as a

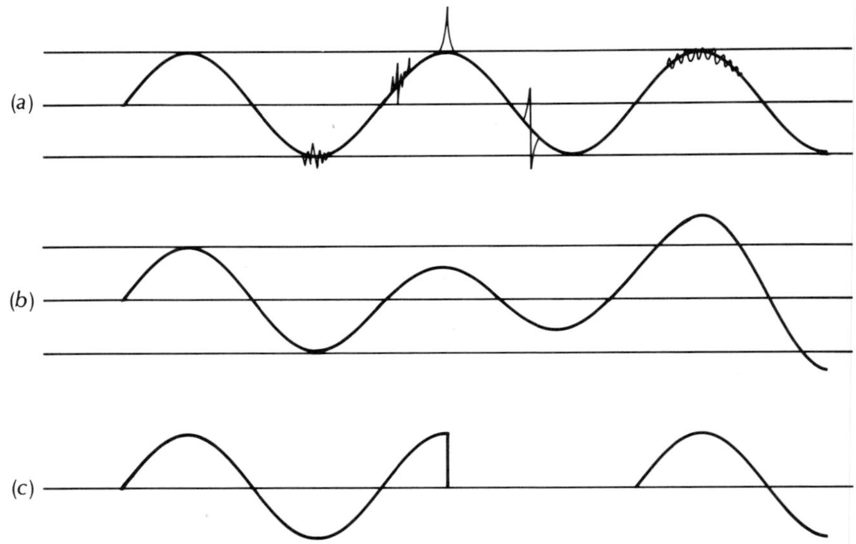

FIGURE 1-3 (*a*) various noise, spike, and ringing oscillations may appear on AC waveform; (*b*) power line sag or surge may last from only a small part of the cycle to all day; (*c*) complete loss of AC can occur from many causes, but is the least common of all.

traveling wave. The traveling wave will travel at near the speed of light back and forth between the point of the fault and the transformer, causing oscillations and voltage swings of twice the normal line voltage. These oscillations and voltage swings are not limited to the faulted line, but will couple over into other lines, as well as into signal wires that may be located near the power wiring. Traveling waves will often be the cause of malfunctions or faulty operation of equipment where the malfunction is often attributed to the AC power, but there is little or no observable indication of a problem in the lighting in the building or operation of less sensitive equipment.

System switching transients and lightning effects are the two major causes of surges in low-voltage AC power circuits. Lightning that strikes either the primary or secondary circuits of the power system will generate surges, but the lightning does not necessarily have to strike the wires to cause a surge. Near-miss strikes that hit an object near the wires can set up electromagnetic fields that induce voltages on the conductors, and ground strikes can enter the common ground impedance paths of the grounding network. There can be major power system switching disturbances, such as capacitor bank switching, or something simpler such as a large piece of equipment within the same building switching off abruptly, or resonating circuits associated with switching devices, such as thyristors, and various system faults, such as short circuits and arcing faults. One switching transient, for example, results from fast-acting current protective devices such as current-limiting fuses and circuit breakers capable of arcing times of less than 2 μs. These devices leave trapped inductive energy in the circuit upstream, and on collapse of the field, high voltages are generated. Transient overvoltages associated with the switching of power factor correction capacitors, on the other hand, have lower frequencies than do the high-frequency spikes with which this book is concerned. Their levels, at least in the case of restrike free switching operations, are generally less than twice normal voltage. Nevertheless, they should not be disregarded. A switching operation involving restrikes is another example. Air contactors or mercury switches can produce, through escalation, surge voltages of complex waveshapes and of amplitudes several times greater than the normal system voltage. The most visible effect is generally found on the load side of the switch and involves the device that is being switched as well as the switching device. In the case of the device being switched, the prime responsibility for protection rests with either the manufacturer or the user of the device in question. The presence and source of transients may be unknown to the users of those devices. This potentially harmful situation occurs often enough to command attention.

The rate of occurrence of surges varies over wide limits, depending on the particular power system. Prediction of the rate for a particular system is difficult and usually impossible. The rate is related to the level of the surges, in that low-level surges are more common than high-level surges. It is essential to recognize that a surge voltage observed in a power system can be either the driving voltage or the voltage limited by the sparkover of some clearance in the system. Therefore, the term "unprotected circuit" must be understood to be a circuit in which no low-voltage protective devices have been installed, but in which clearance sparkover will eventually limit the maximum voltage. The distribution of surge levels, therefore, is influenced by the surge-producing mechanisms as well as by the sparkover level of clearances in the system. This distinction between actual driving voltage and voltage limited by sparkover is particularly important at the interface between outdoor equipment and indoor equipment. Outdoor equipment has generally higher clearances, and hence higher sparkover levels. A level of 10,000 V may be typical, but 20,000 V is possible. In contrast, most indoor wiring devices used in 120-V systems have sparkover levels of about 6000 V. There have been many studies on the rate and magnitude of surges, and they can be broken down as follows:

1. Bell Telephone Laboratories data indicate surges to 1000 V at about two occurrences per year.
2. General Electric reported surges to 1200 V at about one occurrence per year.
3. General Electric data indicate a ratio of 100:1 in the rate of occurrence from 2000 to 6000 V.

Other groups have found similar data, with variations naturally based on geographic location.

Both the lightning and switching disturbances can be in the form of the traveling wave discussed earlier. Even without the presence of lightning, surges can occur several times a day. These disturbances are not limited in duration to a few cycles, but can last for several minutes or more. Power-line fluctuations that cause the lights to flicker are obvious, but not so obvious are the ones that last only 50–100 ms. These short-duration power disturbances can be very destructive and can occur continually. The long term effect can be the degradation of components on the input to expensive equipment and an early death for that equipment. Later in this text there will be a discussion of source impedance. To prevent misunderstanding, a distinction between source impedance and surge impedance is needed. Surge impedance, also called the *characteristic impedance*, is a concept relating the parameters of a line to the propagation of traveling waves. For the wiring practices of the AC power circuits discussed in this text, this characteristic impedance would be in the range of 150–300 Ω, but because the durations of the waves being discussed (20–50 μs) are much longer than the travel times in the wiring systems being considered, traveling-wave analyses are not useful here. Source impedance, defined as the impedance presented by a source of energy to the input terminals of a device, or network, is a more useful concept here. In the conventional Thevenin description, the open-circuit voltage, at the terminals of the network, and the source impedance of the surge source, are sufficient for calculation of the short-circuit current, as well as any current for a specified suppressor impedance.

While surges represent one of the most common and also most destructive types of power-line disturbance, the problems associated with noise riding on the AC waveform must not be overlooked. Noise generated by automobile ignition, transmitters, power tools, and machinery can be picked up external to the building and coupled through the input transformer into sensitive equipment, causing erratic behavior, as shown in Fig. 1-4. Power tools and

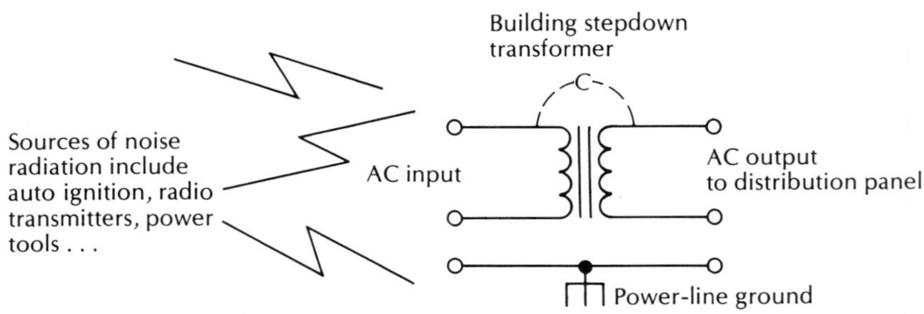

FIGURE 1-4 Winding capacitance *C* will couple noise appearing on both primary leads to secondary leads measured with respect to power ground.

machinery operating within a building can cause a similar problem unless precautions are taken to isolate their sources of power from those of the equipment you are attempting to protect. Electromagnetic interference (EMI) is the unwanted energy in the frequency range from less than 60 Hz to the end of the spectrum, whereas the frequencies of radio frequency interference (RFI) make up only a small portion of the electromagnetic spectrum. Radio frequencies range from about 10 kHz to more than 1000 MHz. The Federal Communications Commission (FCC), effective in 1981, imposed regulations restricting the amount or RFI emitted by "computer devices." These regulations are limited to the RFI portion of the EMI spectrum. The regulations are intended to minimize interference with communications from these sources. The FCC regulation for radiated emissions applies for frequencies above 30 MHz. The regulation distinguished between commercial and household computing devices. The field-strength limits are held to much lower levels for class B (household and residential devices), compared to class A (commercial) requirements, their concern is the proliferation of residential types of computing equipment. Several manufacturers of modern AC power-line conditioning equipment have designed their equipment to meet these same standards imposed on the computer manufacturers by the FCC. Not all equipment, however, will meet the RFI limitations specified by the FCC, and it is possible that some devices will be placed in service in positions that may prove harmful to susceptible equipment. In October 1979 the FCC published in the Federal Register new technical standards governing computing equipment, which apply to all such equipment manufactured after July 1, 1980. The new rules define a computing device as "any electronic device or system that uses digital techniques. More precisely, it is an electronic product that intentionally generates and uses radio frequency energy in excess of 10,000 cycles, (or pulses) per second." The regulation specifically lists business and personal computers, data processing equipment, digital weighing scales, switching power supplies, electronic games, electronic cash registers, digital watches, pocket calculators, and digital clocks as examples, but also notes that the list is not meant to be conclusive.

Filters are designed to reduce the interfering noise signals to or from electronic devices that generate or are susceptible to such noise. They are designed to suppress RFI voltages that exist line-to-line (differential mode), line-to-ground (common mode), or both. Common-mode interference is the most frequently encountered noise problem. The signals are readily propagated in power systems and equipment. Susceptibility problems are generally the result of common-mode interference. Some devices, such as silicon controlled rectifiers (SCRs) and switching regulators, can cause differential-mode interference to nearby susceptible equipment, but this interference is rapidly attenuated in most wiring systems.

Long-term sag or brownout of the power line is another problem that may be the cause of power distribution equipment external to the building, or could be artificially created within the building through the improper loading and distribution of power lines to critical equipment.

As shown earlier (Fig. 1-2), equipment that may operate only for short times or have high initial starting current will load down the power line it is connected to, often affecting the operation of other voltage-sensitive equipment sharing the same line. Figure 1-5 shows an example of a similar installation, but in this case the heavy load operates continually. The length and size of the lines feeding the heavy load will determine how much voltage is dropped when the load is turned on. The problem encountered with this installation is the voltage may be sufficient to operate the heavy load, but is marginal for the critical load. Any further reduction in voltage due to external forces may cause the critical load to fail.

Of all the various types of power disturbance, the one that occurs the least is perhaps the most difficult to deal with. Complete loss of power, blackouts, lasting from a few cycles to several hours, will certainly interrupt the operation of any equipment that is not intentionally protected against power failure. Smaller equipment that may use internal battery backup will not be disturbed, but critical equipment used in factory automation, computers, and similar equipment will definitely be disturbed by power failures lasting only a single cycle. Computers typically have a "power good" detect circuit that will shut down the system in an orderly manner on detecting a power failure. While the power supply may have sufficient capacity to ride through a short outage, the power good circuit will react very quickly and start the shutdown sequence while the power supply can still operate. While complete failure of the power line is the rarest of occurrences, when it does happen the result can be the loss of critical data and time. Some computer and industrial systems may take from hours to days to return to normal operation after a power failure. A number of years ago a liquid fertilizer processing plant had a requirement for an uninterruptible power system (UPS) for its process control. In this instance the plant needed to keep power going to some temperature-

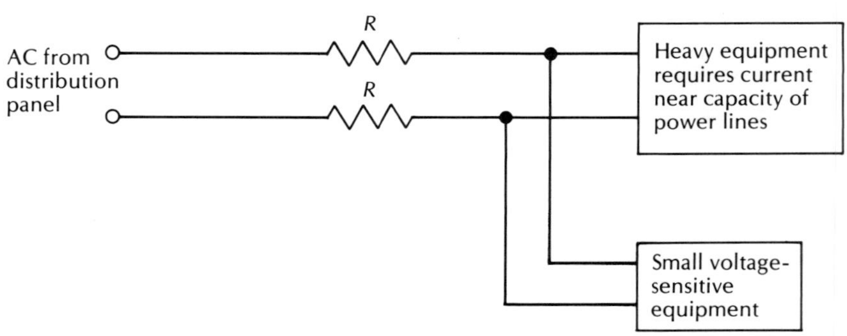

FIGURE 1-5 Line resistance *R* will cause a reduction of voltage to equipment due to high-current requirement of heavy equipment.

sensing systems and automatic-shutdown control circuits. Whenever they would experience even a momentary power interruption, the whole plant would suddenly shut down. This plant was built with all the plumbing outdoors, a real maze of pipes going in all directions. Those pipes carried extremely hot liquids, and when the plant operation would suddenly shut down, all the plumbing would suddenly start twisting and bending, literally tearing itself apart. It would take up to several weeks to get the plant back to operating status after one of these unscheduled outages, and once the UPS was installed the problem disappeared. It was recognized by the plant owners that the cost of the UPS was covered by just one single outage that did not result in total plant shutdown.

It is just as common, if not more so, for the failure to be internal to the building as external. If someone overloads a circuit and trips a breaker in the distribution panel, the result—in terms of everything connected to that line—is the same as if power were lost to the entire city. Many larger industries and hospitals have emergency backup generators that start, sometimes automatically, whenever the main power to the building fails. This still does not protect critical equipment totally. The time it takes for the generator to come on line is longer than most electronic equipment can operate without power, and the problem of internally generated brownout and blackout conditions still exists, even though the source of power is a generator within the building.

Devices have been available for nearly as long as commercial AC power whose intent is to "clean up" the AC power, in one way or another, to reduce the risk of problems encountered by the user. Prior to 1975 or so most people in daily life at home or work found little need for any type of power protection devices unless they were working with specialized equipment of some type that was sensitive to power-line disturbances. After then, all that changed. By the late 1970s computers played an increasingly important role in the workplace, and soon most industrial personnel became accustomed to having computers helping them with their work. Among the first to feel the effect were secretaries, who found word processors to be far superior to any electric typewriters of the time. The ability to quickly insert, delete, modify, and edit text opened up a whole new marketplace for computer manufacturers. Granted, the first word processors were generally only that and have since been displaced somewhat by personal computers with word processing software, but they were none the less computers. At the same time that the world discovered the joy of word processing it also discovered what users of large computer systems had known all along. If you don't have reliable AC power, you risk scrambling your data. Users quickly discovered the importance of saving their work as they went along in order to protect it against sudden power-line failure. That is fairly simple and straightforward, but now computers are becoming more and more a part of the workplace for other uses besides word processing. Smaller businesses that conducted all their accounting by hand were becoming familiar with desktop computers and accounting software that would make their job faster, simpler, and more accurate. The same discoveries were being made by all the other departments within small businesses as well; hence the sudden tremendous growth of the desktop computer market in the early 1980s. Computer-assisted design (CAD) workstations have become as important to designers and drafters as the word processor has become to secretaries. Unlike word processing and some other tasks that are performed with computers, it is not always possible to save your work as you go in order to protect it against sudden power failure. An engineer performing a circuit simulation on a workstation may spend only a few minutes setting up the simulation, but then go away for many hours while the computer crunches the numbers and provides the results to a data file. Once the simulation is complete and the data is safely tucked away in mass storage, a power failure will generally be of little consequence. Murphy's Law being alive and well, however, it is just as likely that a power disturbance will occur while the computer is working, as not. Some simulations can be run overnight or even take the weekend to run. Failures that interrupt this work lead to not only frustration but also lost time and in some cases incorrect data.

One intended use of computers today is to not only assist workers in performing their jobs but often to free them up to do other things while the computer works away unattended. Some process control systems, robots, and even the engineer circuit simulation mentioned above, if left to operate unattended, must have reliable power, or disaster will result. And robots don't necessarily have to look like the classic types seen these days in advertisements shown welding cars together. Your robot may be a simple copy machine that will copy and collate 25 copies of a 100-page document while you are off doing other work. If that in itself is an important enough task that power-line problems may interrupt, then that is your robot doing your work for you and it deserves the same protection as any computer. Power-line disturbances don't always display themselves as hard failures within the computer, or abrupt termination of a software program, but will often be disguised as incorrect data. It is worse if the disturbance occurs while saving data to mass storage. Now you not only run the risk of losing your current work but may also have corrupted the file and scrambled the data previously saved. Networks are good candidates for problems associated with power disturbances. With individual computers linked to one central computer it is often disastrous to the entire network to have one workstation terminate communication without signing off. Another problem is encountered on the cables that link the computers together. High-speed data transmission is taking place, and noise on power cables that are in close proximity to the data lines can cause the computer to misunderstand the meaning of the data, resulting in a failure to properly communicate, or scrambling of the data. Often that scrambled data is saved that way in the system, and you don't discover that your data is gone until next time you attempt to access it.

During a recent trip to Disney's Epcot Center, the author observed the problem of power disturbances and robots graphically demonstrated. An entertaining demonstration of how robots can be programmed to perform various duties normally carried out by humans suddenly resulted in abrupt termination and several hours of work by the maintenance crew to get the show back on. Electrical storms are frequent in Florida, during the summer, and any computer not protected by a UPS will undoubtedly spend a fair amount of its time without power. It turns out that this particular display was being backed up by a UPS system, but the system failed to perform when the power went off. This serves to illustrate two ideas that are vital to the concept of not only power conditioning but also the idea of providing continuous power to a system. First, no matter what action you take, there is always the risk that the very equipment you select to provide you with AC power protection will eventually fail itself. Second, no matter how much money is spent on the power protection equipment, it likely will have a mean time between failure (MTBF) that is shorter than the power line, taken over a year's time, it is intended to protect. This is not necessarily the case with low-cost, simple line filters, or even some types of line regulators. As the degree of sophistication increases, and the AC power protection equipment actually replaces the power line to your load, however—as is the case with many UPS systems—the risk of shorter MTBF is real. Particularly when it is considered that the complete loss of AC power, in many parts of the country, is the rarest of all types of failure. What you must consider is not necessarily the risk of failure of the UPS, but rather the very low odds of having the UPS fail at the same time that the power line has failed. If properly maintained, it will generally be there when it is needed. Most UPS systems are available with a fast bypass switch that will automatically connect the load to the AC line in the event that a failure within the UPS should occur. This will enable the user to make an orderly shutdown of the equipment, when convenient, and service the UPS.

There have been many studies on the quality of AC power in various parts of the country, and the results of those studies all seem to indicate that things are not getting any better. The quality of power is a fairly subjective topic, depending on whom you talk to, and while there are some standards that describe what it is supposed to be, it is difficult for the average person not engaged in the power protection business to relate to the subject. The quality of the power is really only as bad as it causes problems for equipment. Some equipment will not be sensitive at all to the power disturbances that may completely destroy other equipment. Some short dips in the voltage will disrupt the operation of some systems, while others ride right through without noticing. One good test of how reliable your power may be is to consider how many times in the past month, or 6 months, you have had to reset all the digital clocks in the house. If the power goes off long enough to cause the clock in a VCR, microwave oven, or other computerized appliance, to lose its brains, then you can expect that any computer or other similar device will also be disturbed. And that only points out the disturbances that were more severe. Many disturbances in the power line, as will be discussed later on, do not result in a loss of voltage but will cause too much voltage, or make damaging voltage spikes. Only specialized instrumentation can detect these types of disturbances.

When installing sensitive electronic equipment it is often important to take into consideration what type of business takes place at that location. A site survey will often reveal potential problem areas that need to be addressed with some form of power protection equipment. A small electronics company, for example, that manufactures digital meters will not be as likely to have operations going on within the business that would cause internally generated power-line problems as would a business that uses a lot of heavy machinery. Another consideration is the age of the building in which the business is located. Many older buildings may not be wired to current standards that are required for optimal operation of computers or other sensitive equipment. The wiring may be fine for the machinery already in place, but when the owners want to upgrade their accounting systems with computers, then consideration must be given to how they will work in the environment and what sort of power protection may be required. Several companies produce power-line monitoring equipment that is programmable, and some are available with a paper-tape display. This equipment is useful for determining over a given period of time just what is encountered in the way of power-line disturbances, and the fact that most of this type of equipment is programmable, allows the user to tailor it to the equipment that is planned to operate on the line being monitored. In other words, it can be set to ignore events that would not disturb the equipment and to report on events that would. Generally the displayed data includes time of day and a description of the event, (spike, surge, etc.). These monitoring devices can be a valuable tool.

The level of design sophistication found in modern power protection equipment varies greatly, being dependent mostly on the particular market served. Cost is a strong motivator and is likely the single strongest motivation to advancing the state of the art, either to accomplish more with less or introduce technologies into the products that help justify the total cost. As computers have advanced, and microprocessor technology has led to low cost, some power protection products are utilizing this technology. Control circuitry for some line conditioner and line regulator products have included microprocessor control for several years. The same is true of UPS products. For instance, RS-232 communication between a UPS and the computer it is backing up has become a common occurrence. This communication will allow the computer to keep tabs on the power environment and in some cases will allow the computer to make an orderly shutdown of its operating system, without human intervention, when it determines that a failure of the power line has occurred, and has not come back on within a given period of time. The communications link to the power protection equipment is helpful in other ways as well. A service technician can contact the equipment via modem and possibly diagnose trouble, or detect a potential problem developing before it gets too far.

Summary

It is becoming more common to find computer manufacturers offering power protection products as a package with their computer systems, even giving extended warranty when power protection is included in the installation. This acknowledgment by the computer industry of the presence of troublesome power-line-related problems, and their recognition of various types of equipment as a solution to the problems, has added a high degree of credibility to the whole power-line protection industry. It has also created an incentive for many companies to enter the business. All the power-line problems presented here can be harmful to electronic equipment or the data it contains, but all the problems can be minimized by the addition of some form of AC power conditioning. AC power conditioning can take the form of anything as simple as a low-cost surge arrestor to uninterruptible AC power systems that use batteries to create an AC voltage to replace the existing AC power line. In some parts of the country where total loss of the utility power is rare, and the equipment being protected has no need to operate during power failure, less expensive line conditioners or regulators will serve the need. Other areas that experience higher rates of power failure may necessitate the UPS solution. The following chapters will deal with many of the various types of AC power-line conditioning equipment currently in use, including a discussion of how it works and design examples that will assist an engineer in developing these products. It is not the intent of this book to train engineers, but rather to introduce engineers to the field of AC power conversion and provide some insight into the power converters used.

PART II

Noise and Surge Reduction

Noise and surges on power lines are two distinct and different problems, but are often caused by similar sources. In addition, the solution to one will often have an effect of improving the other as well. This section will present the problem of noise and surges on the power line as separate issues, along with solutions to the problem, but the reader should keep in mind that most equipment that is sensitive to one problem will likely encounter difficulty with the other, and both should be considered when power-line protection is being considered.

CHAPTER 2
RFI Noise Reduction

Conducted radio-frequency interference (RFI), energy conducted via the power connection into digital electronic equipment, may cause scrambling of data in computers or other malfunctions. The interference can be in one of two modes: (1) common-mode interference that is present as a voltage from all the power connections to ground or (2) differential-mode interference that is present as a voltage between the power connections.

When RFI susceptibility problems are tracked down, common-mode interference is found to be the cause more often than not. It is distributed within power distribution systems and is also induced onto the power cord by radiated RFI energy and is, in turn, radiated from the wiring into the adjoining space, where it can enter signal wiring.

On the other hand, differential-mode interference is generally found to be a problem only when the radiating source is in close proximity to the susceptible equipment. Building wiring will usually attenuate this mode rather quickly, and the power cord will not receive or radiate it. Since the source of the susceptibility problem could be either mode, it is a good practice to select a design that will attenuate both common- and differential-mode RFI.

Typical power-line RFI filters, as depicted in Fig. 2-1, are passive, bilateral, two-terminal pair networks, with a complex transfer characteristic that can be expressed as

$$T = \ln \frac{P_1}{P_2} \tag{2.1}$$

where P_1 is the maximum power that could be transferred from the generator E to the load R_2, and P_2 is the power that is actually dissipated in the load. Thus the transfer characteristics of the network shown in Fig. 2-2 can be expressed as

$$T_0 = \ln\left(\pm\sqrt{R_1/R_2}\left(\frac{I_1}{I_2}\right)\right) \tag{2.2}$$

$$T_0 = \ln\left(\pm\sqrt{R_1/R_2}\left(\frac{E_1}{E_2}\right)\right) \tag{2.3}$$

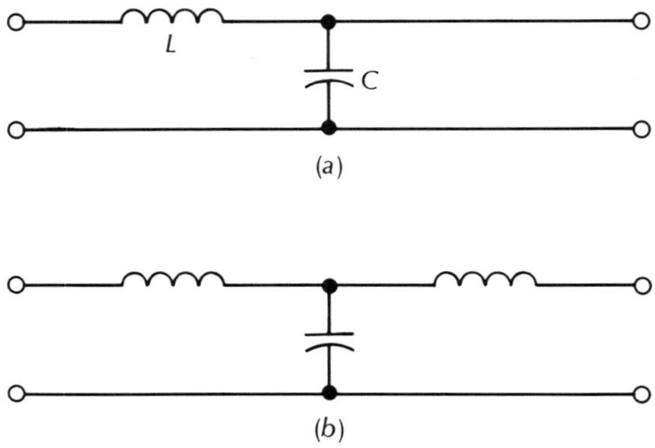

FIGURE 2-1 (*a*) Two-element L filter; (*b*) three-element T filter.

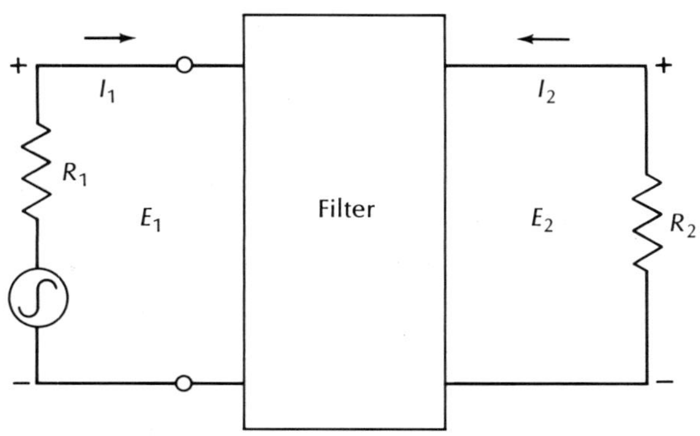

FIGURE 2-2 Filter and associated input and output impedance.

Since the transfer characteristics of the network are functions of the terminating impedances, these impedances must be known if the behavior of the network is to be calculated.

In RFI filter applications, the terminating impedances of the power line and the equipment are not usually known accurately over the important frequency range of 150 kHz to 30 MHz. Therefore, the transfer characteristics of the RFI filter terminated by the equipment and the power line cannot be calculated or derived. RFI filter design is an empirical task, and only past experience will serve as an aid since there is little quantitative data available to help.

Figure 2-3 shows several levels of complexity in designing an RFI filter, and the resultant typical insertion loss for each is shown in Fig. 2-4.

Typical insertion-loss data is generally useless for anything but comparative analysis since it is taken in a 50-Ω test circuit, and neither the power line nor the equipment is likely to present a 50-Ω resistive impedance.

RFI filters can be regarded as mismatching networks at high frequencies. The greater the mismatch is to the terminating impedance, the more effective is the filter in attenuating RF energies. Since the power line is considered to

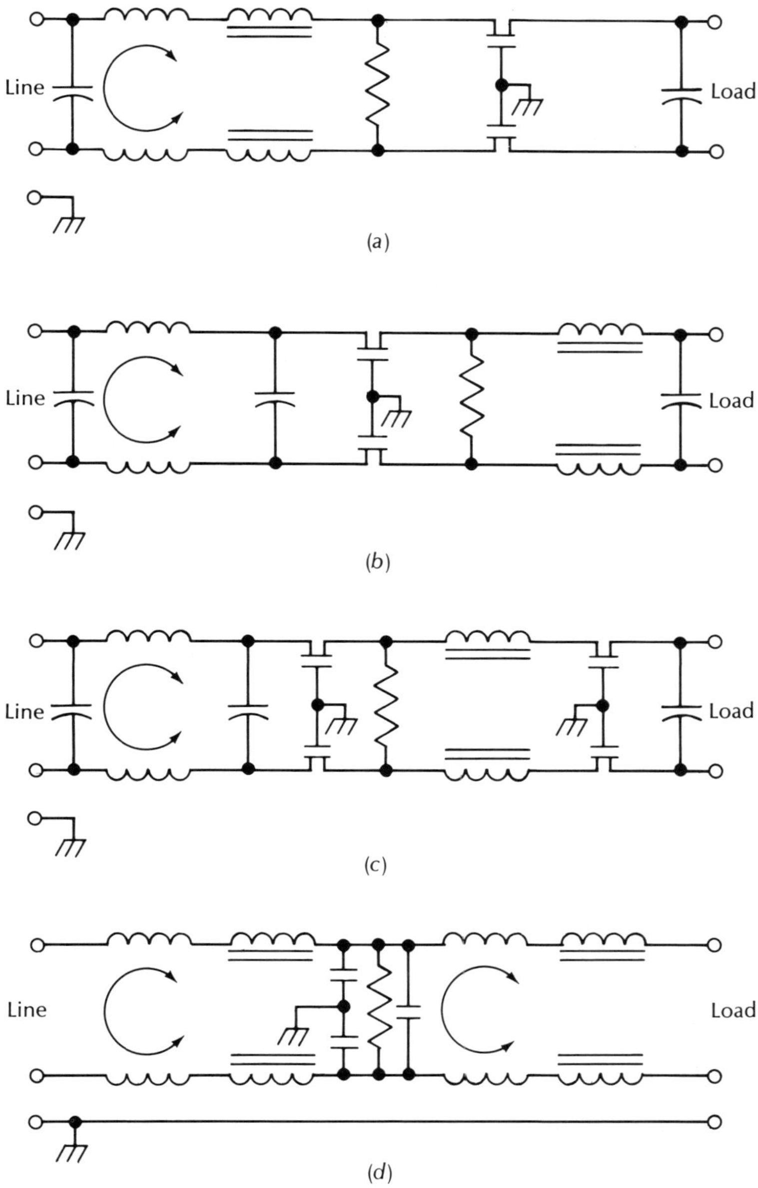

FIGURE 2-3 Filter circuits with increasing degrees of complexity.

be a low impedance, the filter should present to it a high impedance, like a series inductance. If the equipment presents a high impedance, the filter must present a low impedance, like a shunt capacitor. Figure 2-1*a* depicts this as an L filter. If the equipment is determined to be a low impedance at high frequencies, the filter must present a high impedance, and this requires a series inductor, depicted as a T filter in Fig. 2-1*b*.

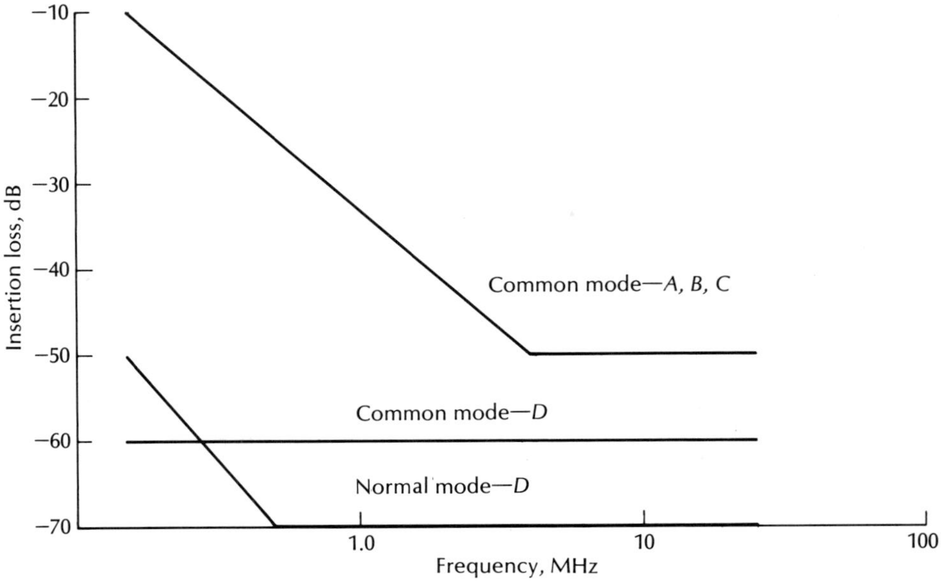

FIGURE 2-4 Insertion loss in 50-Ω system typical per MIL-std-220A.

Several manufacturers of RFI filters offer a standard variety of many different configurations. An attempt to design a filter for equipment that is determined to be susceptible to conducted RFI could be an expensive and frustrating task when quite often an off-the-shelf solution is at hand. The ideal solution to RFI problems is to solve it at the source of the radiation.

To further understand the makeup of a power-line filter, refer to Fig. 2-5. This design will suppress both common-mode and differential-mode conducted emission. A common core balun inductor, L_1 has two windings of equal turns and the polarity as shown. When the polarity of the windings is as shown, equal currents in the line and neutral cancel the flux in the core, thus preventing saturation due to AC line current, and the full inductance is available for common-mode suppression. The common-mode equivalent circuit effectively short-circuits the line and neutral, thus placing the two windings of L_1 in series. The remaining inductors, L_2 and L_3, are wound independently and serve for differential-mode suppression, as are the capacitors C_1 and C_4. All the components contribute to suppression of differential mode, except that the inductance of L_1 is now only the leakage inductance between the windings. Capacitors C_2 and C_3 provide suppression of common-mode emissions. Several considerations need to be made in selecting the value of C_2 and C_3. Capacitors connected from one side of the supply line to earth ground could present a potential shock hazard if the ground line should be open-circuit. Because of this these capacitors are limited in value by international safety standards imposed by IEC and VDE. The current flowing to ground at the power line frequency is determined by $I = E/X_C$ or $I = E(2\pi fC)$. Figure

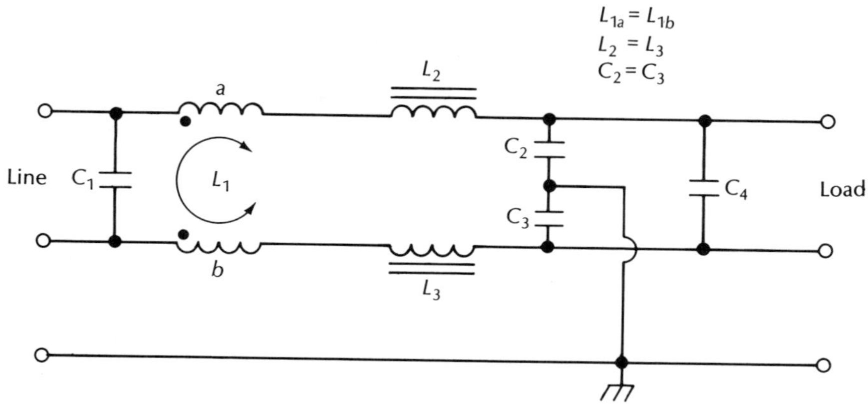

FIGURE 2-5 Power-line filter will suppress both common- and differential-mode conducted emissions.

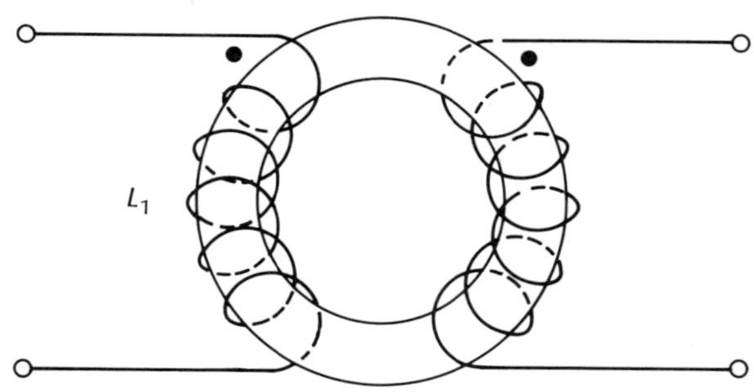

FIGURE 2-6 Physical winding of L_1.

2-6 is an example of how L_1 is physically constructed. Each of the windings must be rated to conduct the required AC current, but since the flux is canceled by this current, a high permeability core material, with little or no air gap, may be used. This allows for a high inductance value while maintaining a small size. Each of the windings has the same turns so bifilar winding makes construction easier, but the bifilar turns increase the interwinding capacitance, and the inductor could resonate at a relatively low frequency. The winding method for the toroid core shown in Fig. 2-6 is the preferred method, where the windings are physically separated to maximize the resonant frequency.

Design Example

Using the same circuit as shown in Fig. 2-5, assume a 120-V, 60-Hz power line. First, the values of C_2 and C_3 are calculated. Ground current is to be limited to 0.5 mA.

$$X_C = R = \frac{E}{I} = \frac{120}{0.5 \times 10^{-3}} \tag{2.4}$$

$$X_C = 240{,}000\ \Omega$$

since

$$X_C = \frac{1}{2\pi f C} \tag{2.5}$$

then

$$C = \frac{I}{2\pi f E} = 0.01\ \mu\text{F} \tag{2.6}$$

If the common mode cutoff frequency is selected as 18 kHz, then

$$f_C = \frac{1}{2\pi\sqrt{LC}} \tag{2.7}$$

Allowing that C_2 and C_3 are parallel in common mode, and solving for L

$$L_1 = \frac{\left[\dfrac{(18{,}000)^{-1}}{6.28}\right]^2}{0.02 \times 10^{-6}} = 3.9 \times 10^{-3} \text{ or } 4 \text{ mH} \tag{2.8}$$

Since, as stated earlier, the two windings of L_1 are effectively placed in series for common mode, each will have a value of 2 mH, and the contribution of L_2 and L_3 can be considered negligible at this point. Capacitor C_4 is selected

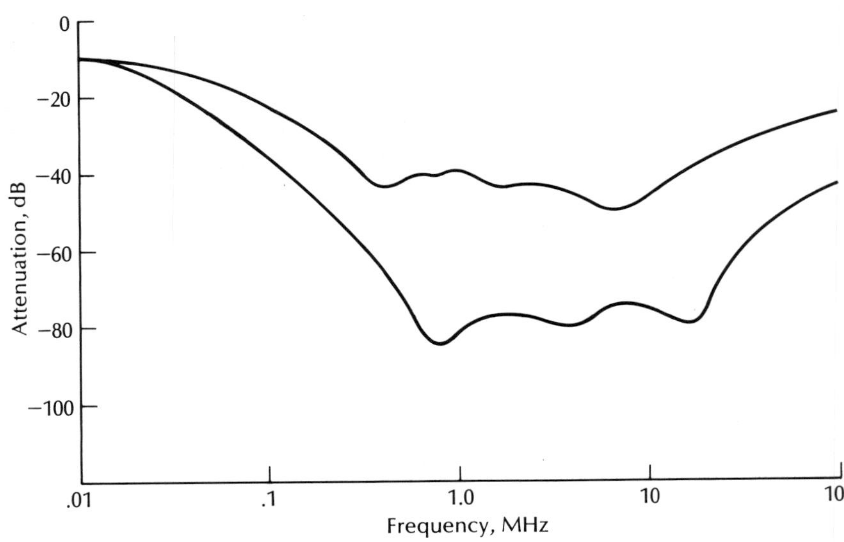

FIGURE 2-7 Insertion loss for filter in Fig. 2-5.

to be 1 μF as a convenient value, and the differential-mode cutoff frequency is selected to be 16 kHz. Solving again for L:

$$L_2 = L_3 = \frac{\left[\frac{(16{,}000)^{-1}}{6.28}\right]^2}{1.0 \times 10^{-6}} = 99 \times 10^{-6} \text{ or } 100\ \mu\text{H} \tag{2.9}$$

The total 100 μH is divided between L_2 and L_3 as 50 μH each. Finally th input capacitor C_1 is selected as 0.05 μF. Given a source impedance of 50 Ω and a load impedance of 100 Ω, the common-mode and differential-mode attenuation characteristic for this circuit is shown in Fig. 2-7.

A discussion of isolation transformers later in this section will provide another method for reduction of high-frequency noise.

CHAPTER 3

Surge Reduction

Surges can vary in amplitude and duration in any power system. All power lines have inductance, and transient voltages will appear whenever the line carrying current is interrupted. The speed at which the current is interrupted, as well as the inductance of the line, will determine the magnitude of the voltage transient, as given in the equation $e = -L\, di/dt$. Studies have shown the waveform in Fig. 3-1 to be typical of what can be found in indoor installations. The natural resonant frequencies of the conductor system will result in oscillatory surges of different amplitudes and waveshapes at different places in the system. These oscillatory frequencies of surges range from 5 kHz to more than 500 kHz.

The more typical surge, as shown in Fig. 3-1, is a 0.5-μs, 100-kHz ring wave. This waveshape rises from 10 to 90% in 0.5 μs, then decays while oscillating at 100 kHz, each peak being 60% of the preceding peak. This fast rise time can produce the effects associated with nonlinear voltage distribution in windings and the dv/dt effects on semiconductors. Installations near the service entrance are in considerable more danger as a substantial amount of current is available. Here the oscillatory waveform is of less interest, and the waveforms shown in Fig. 3-2 will apply. The rise time to peak voltage is slightly longer, but the voltage remains at a high level for a much longer period of time than in the ring waveform. Figure 3-2*a* shows the standard waveshape for the open-circuit voltage or voltage applied to a high-impedance device, and Fig. 3-2*b* shows the discharge current or current in a low-impedance device. The energy involved in the interaction of a power system with a surge source, and a surge suppressor, will divide between the source and the suppressor in accordance with the characteristics of the two impedances. In a spark-gap type of suppressor, the low impedance of the arc after sparkover forces most of the energy to be dissipated elsewhere, such as in a current-limiting resistor that has been added in series with the gap. In an energy absorber suppressor a substantial share of the surge energy is dissipated in the suppressor, but its clamping action does not involve the power follow energy resulting from the short-circuit action of a gap. If a surge suppression device is to be effective, it is essential that a realistic assumption be made about the source impedance of the surge. The degree to which source impedance is important depends largely on the types of surge suppressor used. The surge suppressors must be able to withstand the current passed through them by the surge source.

As with RFI/EMI problems, it is often found that equipment within the immediate area may be the source of destructive surges, and the first approach will be to go to the source of the problem for corrective action. When that is neither possible nor practical, then the next solution is to place a surge protection device of some sort in series with the AC lines to the equipment that is to be protected. Spark-gap devices, metal oxide varistors, silicon avalanche, and isolation transformers are four possible choices.

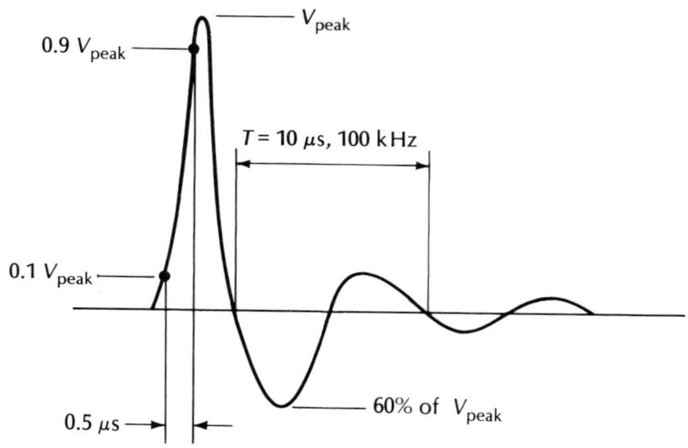

FIGURE 3-1 Typical surge waveform.

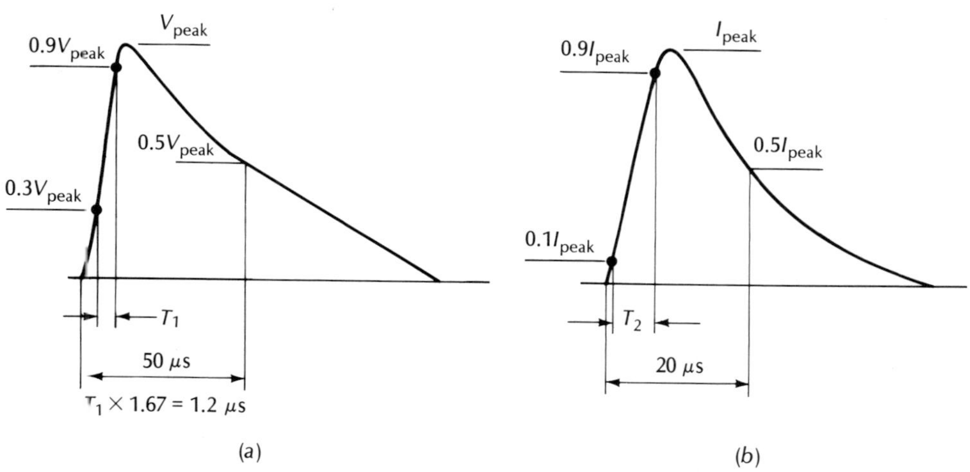

FIGURE 3-2 (*a*) Open-circuit voltage waveform; (*b*) discharge current waveform.

Spark Gap

Spark-gap devices, or *gas tubes* as they are also known, represent an extension to the already built-in protection available at the wall outlet, which is, specifically, the uncontrolled flashover that occurs over or around insulation between terminals in air. As discussed earlier, this occurrence is the limiting factor in most cases where substantially higher voltage would have been experienced. Gas tube suppressors are made up of carefully shaped electrodes of the proper material in an envelope filled with a suitable gas under pressure, which reduces the breakdown voltage so that large electrode spacings may be used, implying better dimensional control and hence greater breakdown voltage stability. By design, the gas-tube breakdown will occur from around 100 V to several thousand volts, for slowly rising applied potentials. This points to the three crucial gas-tube properties that must be understood to apply the device properly. First, they are crowbar devices; when they fire they go into an arc discharge mode, the voltage across them dropping to around 20 V, and the action can occur in just a few nanoseconds. Second, their breakdown voltage is a function of the slope of the applied voltage edge. Finally, the voltage at which breakdown occurs is statistical with various limits from 50 to 300 V, depending on the manufacturer.

The crowbar action of gas-tube devices implies that gas tubes can handle large surge currents of 20,000 A and more in a single shot. It also means that if they are protecting a line whose normal potential is over 10 or 15 V and that has some current capability, they are dangerous to use without inclusion of a specific circuit means for extinguishing the arc. Otherwise, they may continue to conduct after the surge, an effect referred to as *holdover*. At the least, this can clamp and distort the normal signal; at worst, the tube can be destroyed by the follow current if the tube is in a circuit with heavy power delivering capability. The holdover effect will only last up to the end of the half cycle on the AC line, but must be taken into consideration for design with gas-tube suppressors. The tubes sensitivity to applied voltage slope means that a 90-V tube, which might be guaranteed to fire between 90 and 120 V with DC (considered for practical purposes to be a slope of 100 V/s) may crowbar as high as 800–1000 V when a steep edge such as 10 kV/μs is applied. Thus the protection a gas tube furnishes may be adequate for spike surges with low speed edges, yet inadequate for waves with steeper fronts.

Figure 3-3 shows the crowbar action that occurs when a gas tube fires. As the voltage increases from zero, the tube fires when its DC firing voltage is reached. Initially the tube voltage reduces to its glow voltage. As the current increases, the tube drops into its arc voltage mode. As long as the applied voltage is high enough to sustain the arc, the tube remains conductive. Eventually the voltage drops to a level at which the current is insufficient to sustain conductivity. The tube then self-restores and is ready for another event.

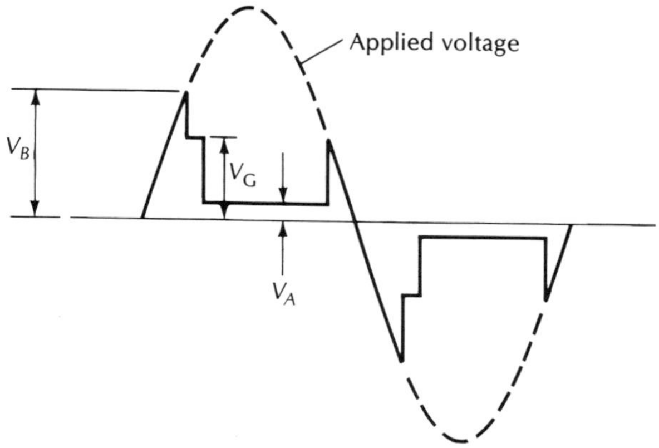

FIGURE 3-3 Voltage characteristics of a gas-filled surge arrester (V_B = breakdown voltage; V_G = glow voltage; V_A = arc voltage).

Metal Oxide Varistor

A metal oxide varistor (MOV) is a nonlinear variable-impedance device consisting of bulk metal oxide particles, separated by an oxide film for insulation. As the applied voltage increases, the film becomes conductive and current flows through the device. Since MOVs clamp, they do not have the current-handling capability of gas tubes for comparably sized devices. They must absorb the transient energy, instead of reflecting it as does a device with crowbar action. On the other hand, ignoring problems of lead inductance, which are common to all protectors, varistors are quite fast and don't have the leading-edge sensitivity or statistical performance characteristics of gas tubes.

The energy dissipated in varistors can be approximated as

$$E = KV_{\text{peak}}I_{\text{peak}}t \tag{3.1}$$

where V_{peak} and I_{peak} are the maximum peak voltage and current, t is the pulse width, and K is a constant as given in Table 3-1. Alpha is a figure of merit and, depending on the size and design parameters of the MOV, will generally fall between 25 and 40.

Design Example

An MOV, as rated from its specification sheet, shows that the maximum voltage across its terminals at 100 A is 520 V. The waveform in Fig. 3-4 is applied, and you must calculate the total energy absorbed by the MOV. Each section of the waveform must be treated separately, so that in this example the waveform is divided into two parts: (1) 0–5 μs and (2) 5–50 μs. From Table 3-1 the value of K will be 0.5 for the rising part of the waveform and 1.38 for the falling portion.

1. $E = (0.5)(520)(100)(5\ \mu s) = 0.13\ J$
2. $E = (1.38)(520)(100)(45\ \mu s) = 3.23\ J$

The total energy absorbed over the full duration of the surge pulse therefore is 3.36 J. It is important that the selected MOV be sized to handle the energy it is expected to have to absorb.

TABLE 3-1
k Based on Alpha Value of 25–40

Wave shape	Equation	K (I)	K (E)
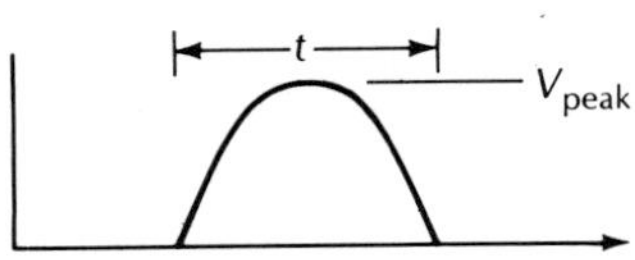	$V_{peak}\sin\left(\frac{\pi}{t}\right)t$	0.63	0.15
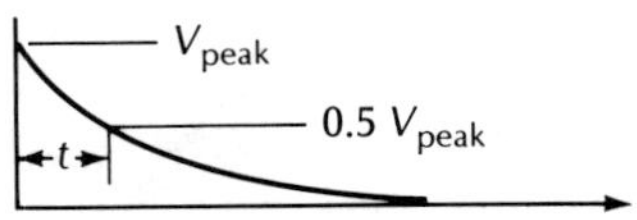	$V_{peak}\ e^{-T/1.4t}$	1.38	0.056
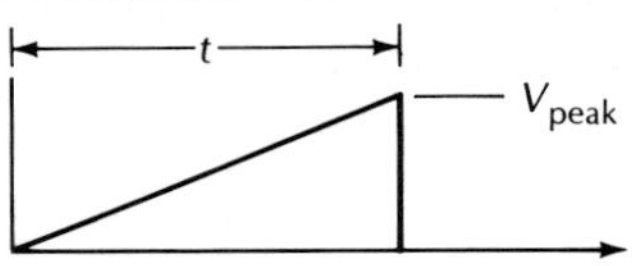	$V_{peak}\left(\frac{\pi}{t}\right)$	0.5	0.038
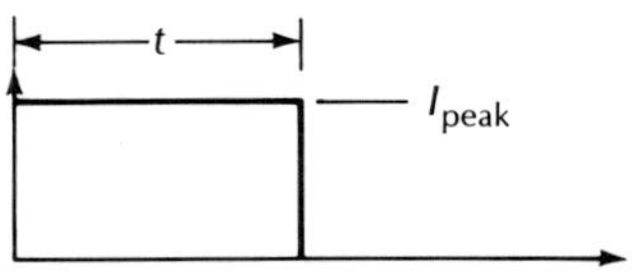	I_{peak}	1.0	—

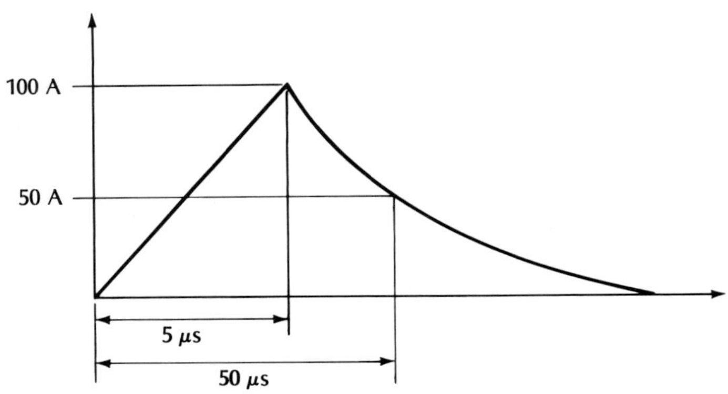

FIGURE 3-4 Transient waveform applied to MOV.

Since the clamping ability of the MOV is of interest here, only the voltages or transients that may occur over and above the rated peak line voltage are to be considered. The worst-case ratio of peak suppressed voltage at a given current to the maximum rated steady-state peak circuit voltage is the effective clamping ratio. For an individual MOV the clamping ratio is solely a function of alpha and the current value. Consulting the individual manufacturer's data sheets will point out the characteristics necessary to making a selection.

The low cost and availability of MOV devices has made them a good choice for surge protection. How they are actually implemented will vary greatly with the requirement. In a power-line surge protector, for example, they will often be used with other passive devices to form a very effective suppressor. This will be discussed later. As a stand-alone suppressor the MOV is also quite effective. If placed right at the input to some piece of electronic equipment the MOV will provide effective protection. In some cases it may be practical to solder the leads of a radial lead MOV directly to the wires where the power cord makes its connection in the enclosure. For rapid clamping action, the shorter the wires to the MOV the better. The inductance in a straight wire is generally small and insignificant. Given a value of 1 μH for a length of 3 ft, most lead wires on suppressors would have an inductance value in the nanohenry range. The voltage drop developed across an inductor under pulse conditions is expressed as $V(t) = L(di/dt)$, where L is in henrys and di/dt is the rate of change of current. For the fast rise times associated with some types of transient the problem is clear, but for slower rise times closer examination is necessary. Tests of the effectiveness of suppressors having various lead lengths at pulse currents ranging from 100 to 400 A, show the importance of keeping the leads short. A 100-A pulse with a waveform of 1.2×50 μs will cause the clamping voltage of the device to increase by approximately 40 V, when the device is placed 24 in (lead length) away from the line it is protecting. This additive peak clamping voltage increases to approximately 100 V at 72 in. A 400-A pulse will add 250 V with 48-in lead wires. It can be seen from this that mounting a suppressor to a circuit board out of convenience, and then placing that circuit board in a position that requires several feet of wire to connect the device across the power line will greatly defeat its purpose.

Silicon Suppressor

A third type of protector is the silicon avalanche suppressor. These devices are extremely large junction zener diodes that have been specifically designed for transient suppression. An attempt at a design that substitutes even a 10- or

50-W ordinary zener regulator for a transient rated silicon avalanche suppressor will surely result in disaster. The two devices are manufactured in vastly different ways. In a spike surge suppression application, a $\frac{1}{4}$-in × $\frac{1}{2}$-inch package silicon avalanche suppressor will exhibit superior heat-sink ability over a power zener. The silicon avalanche suppressor is extremely fast, having a response measurable in the low nanoseconds. Its clamping potential is obtainable by design over a wide range suitable for different applications, from 5 or 6 V to several hundred volts. The lower-voltage versions are ideal for protecting logic circuits of various kinds. As with the other types of device described above, it is important to maintain short leads between the protection device and the power line it is intended to protect.

As with all protection devices, in order to properly select the device, some measurement or best guess must first be made about the nature of the transient that is to be protected against. Source impedance, peak voltage, peak current, and the waveshape should be established. The maximum steady-state voltage rating of the equipment to be protected must then be determined. This is often known as the *standoff* voltage. The repetition rate of the transient and maximum power dissipation the device must handle are critical in selecting the suppressor. Finally the clamping voltage, or the voltage appearing across the suppressor due to peak current and temperature, must be determined. The clamping voltage must be below the maximum operating voltage of the device the suppressor is protecting.

Design Example

It is desired to use a single silicon suppressor, placed across the output of a bridge rectifier operating directly from the power line, to suppress spikes arriving from the power line. The following selection process is offered as a typical process to be followed, resulting in direct selection of a suppressor from the manufacturer's data book.

1. Determine the maximum DC plus ripple voltage on the output of the bridge, plus maximum tolerance.

2. Select a suppressor to have a reverse standoff voltage equal to or greater than the maximum circuit voltage, as defined in step 1 above. This selection will allow for operation over the temperature range of -65 to $+175$°C.

3. Define the waveshape or duration of the transient.

4. Determine the peak pulse current the suppressor will be required to pass. Check this against the maximum peak pulse current shown on the device

data sheet. If the predicted current exceeds the device rating, all is not lost. Something can be done, as will be discussed later.

5. If the current to be handled is within the device rating, proceed to determine the maximum peak pulse power the device must handle. This value is obtained by multiplying the transient peak pulse current by the maximum clamping voltage of the selected device.

6. If peak pulse power is within the maximum rating of the device, use the device selected. To evaluate this, the follow criteria should be used:

   a. For a transient pulse width of an exponential delay, define "pulse time" as the duration of the pulse from the 10% amplitude point on the leading edge to the 50% amplitude point of the trailing edge. From the device data sheet determine the peak pulse power rating of the device for the computed pulse time of the incident pulse.

   b. For transients that approximate a sine wave in pulse waveshape, use the total pulse width as the pulse time and multiply the corresponding peak pulse power value by 0.75 to obtain the effective device rating.

   c. For square-wave-shaped transients, multiply the peak pulse power value for the corresponding pulse time by 0.66 to obtain the effective device rating.

   d. If the transient is a rapidly damped sine wave, or rapidly damped square wave, with one time constant within eight cyclic, rate the same as if the device were subjected to just one cycle of the highest amplitude in the wave train.

   e. For any other transient conditions, consideration must be given to the average power-handling capability of the device, and/or the total transient energy must be summed and evaluated in respect to thermal derating.

7. If the peak pulse power of the incident pulse is greater than the rating of the device, then several devices may be stacked in series to increase the power rating. When stacking the same power-rated devices, the current rating increases in proportion to the number of devices stacked. If it is impossible to achieve the necessary power rating by stacking the devices in series, parallel stacking can be done effectively for devices rated below 100 V. Close matching, about 200 mV between each device, is necessary to assure even loading of the transient between all the devices. Series–parallel combination stacking can be done for higher voltages and power.

Combination Suppressor Circuits

Combinations of the elements described above can be used to make up a protector that will respond to several different types of line disturbances. Figure 3-5 shows such a circuit. In this case a gas tube has been combined with a silicon clamping device to give a two-stage suppressor. When a surge hits, the gas tube will fire and crowbar the bulk of the wave, while the silicon clamping device, being very fast, will catch the leading edge of the wave the tube didn't catch. It is important here to make certain the gas tube and silicon device are sized properly to work together. If the tube doesn't have enough remaining voltage to ensure that it fires, then the silicon device will have to take the full surge energy and may be destroyed in the process.

The series impedance shown in Fig. 3-5 can be a resistor or an inductor, or a combination *RL*. It is sized to pass the full load current while still providing enough impedance to ensure that enough of the leading edge of the surge wave remains to fire the tube. Selecting this series impedance can be a problem since, although there are standard surge waveforms to use for reference, the actual waveform and magnitude of a surge at a given location can only be guessed at. If not properly designed and thoroughly tested, this type of suppression device will leave holes in the protection. That is to say, there will be some wavefronts that will get through.

This combination device has been found effective for suppressing lightning induced transients in systems where data communication between computers and/or peripheral devices was being effected. Lightning activity appears to occur with the greatest frequency in the Southeast, with activity in the Midwest reduced to about half that in the East, and activity on the West Coast is about 10% of that in the Southeast. Peak lightning currents can range from a few thousand amperes to more than 150,000 A with the mean value at about 20,000 A. It is not necessary to have a direct strike, electromagnetic coupling will induce very real and very high voltage transients into the power lines, and it can happen for very great distances from the actual strike. Lightning

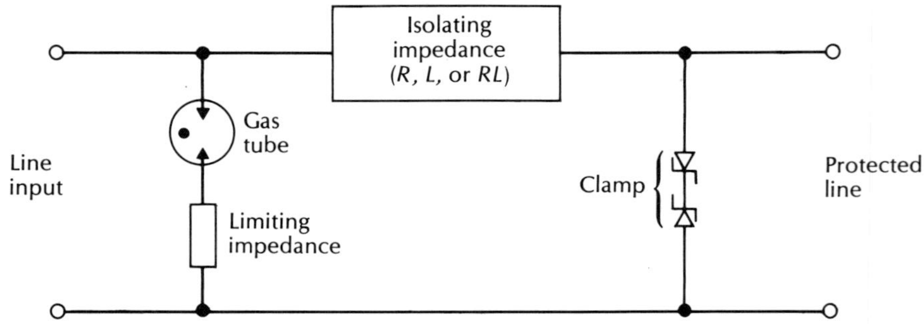

FIGURE 3-5 Two-stage hybrid protector.

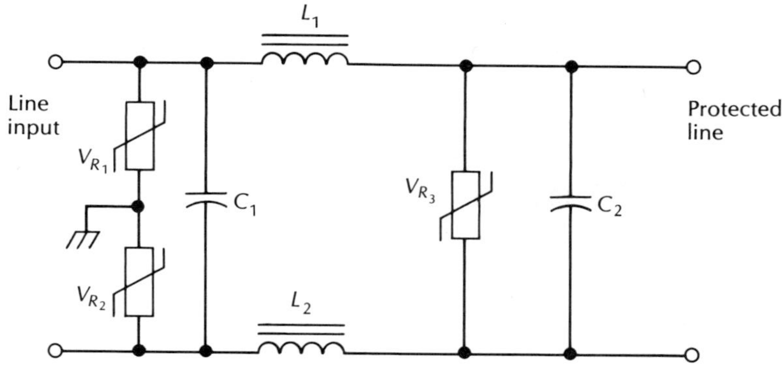

FIGURE 3-6 Multistage protector utilizing MOVs, inductors, and capacitors.

strikes as much as 8–10 mi from a power line can induce transients of several hundred volts in the line. It becomes critical to suppress these voltages before they can reach sensitive equipment and do damage. Often it is not enough to place surge or transient limiting devices ahead of the equipment and expect that to suffice. Sometimes it is necessary to add additional protection right into the sensitive circuitry being protected.

Figure 3-6 is another combination protector. Although it uses only one type of surge device, the MOV, it does combine other circuit elements to provide multistage protection. In this arrangement the MOVs closest to the source of the surge, V_{R_1} and V_{R_2}, will begin to clamp very quickly on the leading edge of the wave. The remaining MOV, V_{R_3}, and capacitor, C_2, working against the impedance provided by L_1 and L_2, will serve as a secondary surge reduction circuit. The response time of the MOV is greatly affected by inductance in series with the leads of the MOV. For this reason it is important that the lead length of the MOV be kept to a minimum. This is particularly important for V_{R_1} and V_{R_2} since they are depended on to reduce the leading edge of the wave, and therefore their response time will be of critical importance in minimizing the peak voltage that gets through. Inductors L_1 and L_2 will facilitate reduction of the rate of rise of the surge current so that V_{R_3} can do its job. A further benefit of this type of circuit is the presence of the capacitors and inductors, which in addition to helping reduce surges, will also provide common- and differential-mode noise suppression as described earlier in this section.

CHAPTER 4
Line Isolation

Power transformers, as found at the service entrance to a building, are designed to operate from nearly zero source impedance, at only the power-line frequency to which it is attached, for the sole purpose of transferring power from the utility. They are isolation transformers from the standpoint that there is no DC path from the input to the output, but unless they are designed specifically to provide a high degree of noise isolation, they will provide little to no attenuation to the common-mode noise discussed earlier in this book. Construction techniques can also be used to provide a means for attenuation of differential mode, which is helpful for suppressing fast rising surge voltages. It is not the purpose of this book to attempt to teach all the finer points of transformer design, as there are many fine books available for that, but rather to present some transformer basics and provide design and construction techniques that will result in a transformer that provides the high degree of isolation needed to perform the tasks of the other devices described in the preceding chapters.

One major drawback to using an isolation transformer, over the other devices described thus far, is its size and weight. But considering that only the transformer can provide DC isolation, as well as an efficient means of stepping voltage up or down, it may be the best solution overall when a transformer is required somewhere in the equipment anyway. The discussions that follow are intended for use with *EI*-type lamination material and single-phase design, although the same methods can be applied to three-phase designs.

Figure 4-1 is a diagram of a simple isolation transformer and its associated equivalent circuit. In this instance we have two copper coils physically placed so that a magnetic field in one links the other through the ferromagnetic core material. When the primary coil is connected to a source of alternating voltage, and the other coil, the secondary, is connected to a load, the alternating voltage in the primary coil produces a changing magnetic field common to both coils and results in the transfer of electric energy from the input circuit to the output. This action is described by Faraday's law:

$$E = -\frac{d(N\phi)}{dt}\,10^{-8} \qquad (4.1)$$

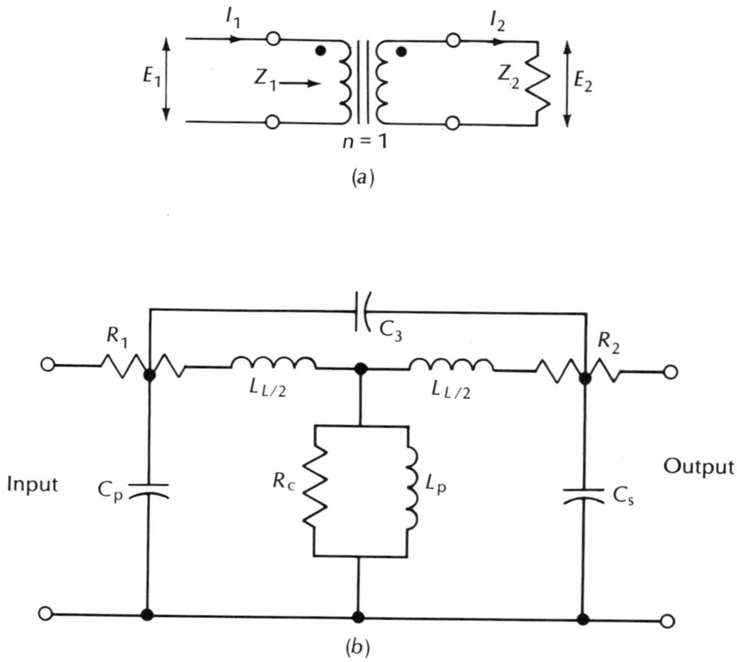

FIGURE 4-1 Transformer diagrams: (a) elementary schematic representation; (b) T equivalent circuit, where $n = 1$, R_1 = primary resistance, R_2 = secondary resistance, R_c = shunt core loss resistance, L_P = primary shunt or mutual inductance, L_l = leakage inductance, C_p = primary distributed capacitance, C_s = secondary distributed capacitance, C_3 = input-to-output leakage capacitance.

where E is in volts, N is the number of turns, and ϕ is the flux. The number of turns in each coil determines the voltages induced in each of the coils:

$$\frac{E_1}{E_2} = \frac{N_1}{N_2} = n \tag{4.2}$$

Current supplied to the primary by the generator is supplied to the load in the secondary, and is inversely proportional to the ratio of the primary to secondary turns:

$$\frac{I_2}{I_1} = \frac{N_1}{N_2} = n \tag{4.3}$$

In an ideal transformer where there is no loss of energy in the form of heat, and there is perfect magnetic coupling, the output power will equal the input power:

$$E_1 I_1 = E_2 I_2 \tag{4.4}$$

Given these relationships, and the fact that the number of turns on the transformer can be easily adjusted to accomplish the desired task, the number of alterations of voltage and current is virtually unlimited.

If a linear relationship is assumed between the flux and magnetizing current, then the expressions for flux density B and magnetizing force H are realized:

$$B = \frac{\phi}{A} \tag{4.5}$$

$$H = \frac{4\pi}{10} \frac{Nim}{l} \tag{4.6}$$

where A is the cross-sectional area of the core material.

The B–H relationship is graphically shown in Fig. 4-2. While all ferromagnetic transformers are nonlinear, if it is assumed for the moment that a linear core material is used, then the resulting straight-line B–H line as shown in Fig. 4-2*a* is realized. In actual ferromagnetic core materials, B and H are not linearly related. The graphical plot shown in Fig. 4-2*b* exhibits, in addition to saturation, a hysteresis characteristic that results in a family of concentric loops. At any given value of H, the value of B depends on the prior history of the core material, that is, on the excitations that have preceded the one under consideration. In addition, any variation in ambient temperature or frequency of excitation will affect the area and slope of the entire family of curves. The complicated relationship between B and H requires a combination of graphical and other mathematical techniques to predict the exact behavior. Since the core material is truly nonlinear, though, it is necessary to set the primary voltage to the transformer so that its maximum never causes the flux density to enter the saturation range, but rather stays within the linear range as shown in Fig. 4-2*c*. All of the previous equations describing the relationship of primary to secondary voltage and current are valid when the transformer is operated in this manner. For a sine-wave voltage as typical to the power line

$$E_{\text{RMS}} = 4.44 B_{\max} f N A \times 10^{-8} \tag{4.7}$$

shows the relationship between root-mean-square (RMS) voltage E, line frequency f, maximum operating flux density $B_{\max}$, and the cross-sectional area of the core A. Examination of this relationship shows that if the input voltage is allowed to increase, the operating flux density will also increase. So, while there are some definite constraints placed on the range of voltage and frequency, for a given transformer design, it can be seen that operated within the constraints the transformer becomes nearly a linear component and can be regarded as such.

Given all the preceding relationships, it is seen that as the input voltage RMS value fluctuates, so does the RMS value of the secondary voltage. Ad-

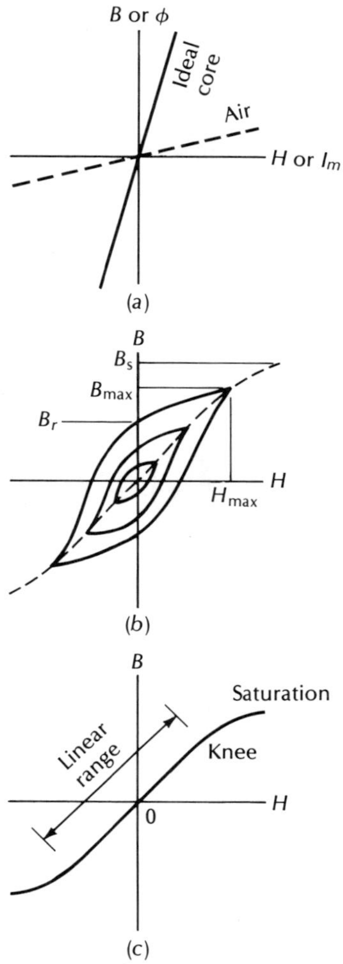

FIGURE 4-2 (*a*) Ideal linear core with high constant slope and no memory B_r; (*b*) a family of hysteresis curves for soft ferromagnetic core material; (*c*) a single-valued curve with nonlinear knee and saturation regions; hysteresis is neglected.

ditionally, any disturbances in the shape of the waveform, within reason, such as what would be found when fast surges or high-frequency noise rides on top of the waveform, will be passed right on through to the load. First, let us look at common-mode noise and how it gets through to the load.

As seen in Fig. 4-3, common-mode noise is generated between either of the power lines and ground. The capacitor shown between the primary and secondary coils is present in all transformers and is called *coupling capacitance*. The presence of coupling capacitance provides an AC path for common-mode noise to pass directly to the secondary coil and the load. This coupling capacitance can be reduced by the addition of shielding between the coils. In Fig. 4-4 a simple Faraday electrostatic shield has been added between the primary and secondary windings. As a result, C is greatly reduced, and C_1 shunts current directly to ground through the shield. This attenuates current through C_2 and the secondary winding, and can reduce common-mode volt-

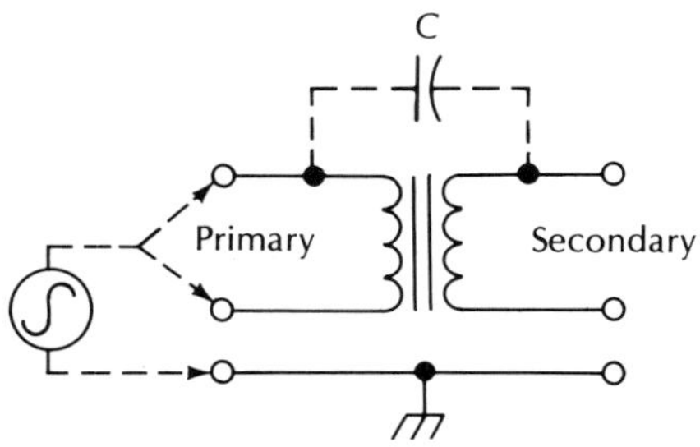

FIGURE 4-3 Capacitor C provides direct path for common-mode noise from primary to secondary.

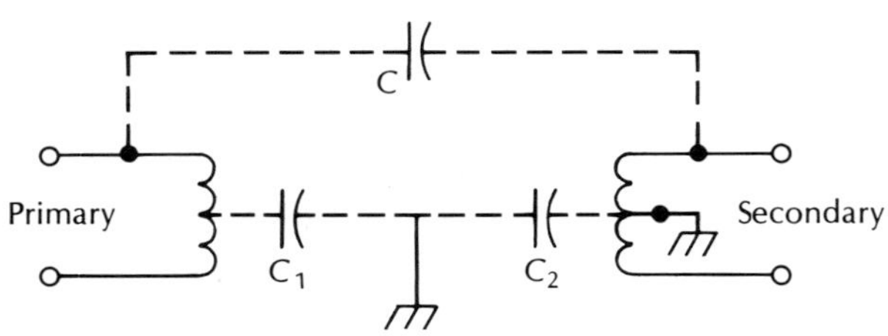

FIGURE 4-4 C is small because of the effective increase in distance between primary and secondary.

age effects in the secondary by as much as 50 dB, when compared with the unshielded transformer. The ratio of input to output voltage, expressed in decibels is given by

$$\mathrm{dB} = 20 \log_{10} \frac{V_1}{V_2} \tag{4.8}$$

from which is derived

$$10^{\mathrm{dB}/20} = \frac{V_1}{V_2} \tag{4.9}$$

Calculating the ratio of output voltage to input voltage, given 50 dB of attenuation, gives a ratio of 316.23. Common-mode transients of greater than 2000 V can be expected on any power line. With a ratio of 316 : 1, a 2000-V transient would appear as 6.3 V on the secondary. This degree of isolation may not be adequate for many installations. In some cases it is desirable to achieve 100 dB or more input to output isolation.

The shielding method shown in Fig. 4-5 is capable of achieving common-mode isolation exceeding 140 dB, or 10,000,000:1. By enclosing both the primary and secondary windings in box shields, and including a Faraday shield between the two, maximum isolation can be achieved. To obtain the best effect, the primary shield must return to the utility ground, the Faraday shield connects to a separate earth ground, and the secondary shield connects to the load ground. It is important to apply some care at this point. If this sort of isolation is really necessary, consideration must be given to the fact that there is now no direct path between the utility green wire ground and the

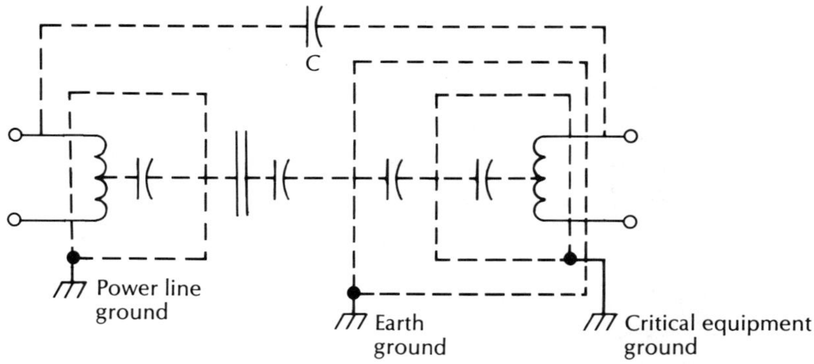

FIGURE 4-5 Isolation transformer with two box shields and Faraday shield can yield greater than 120 dB of common-mode attenuation.

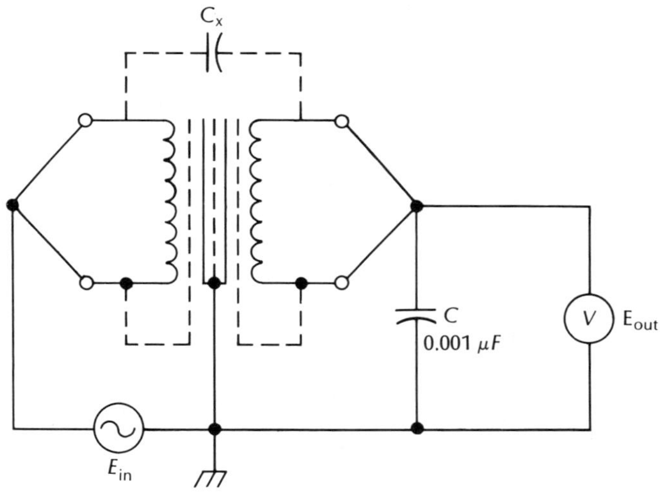

FIGURE 4-6 Measurement method for determining coupling capacitance C_x.

load ground. Safety must be considered at this point, and consideration given to what is given up in isolation in order to ensure the integrity of the ground system. Figure 4-6 is an example of a test setup for measuring the actual common-mode attenuation of a given isolation transformer.

$$V_c = \frac{Z_c E_{in}}{Z_T} \tag{4.10}$$

Therefore:

$$V_c = \frac{C_x E_{in}}{C + C_x} \tag{4.11}$$

$$C_x = \frac{(C + C_x) V_c}{E_{in}} \tag{4.12}$$

Dropping V_c from (4-11) provides

$$C_x = \frac{CV_c}{E_{in}} \tag{4.13}$$

When dealing with transformers displaying 100 dB or more isolation, it is necessary to provide enough source voltage to ensure an accurate reading of the secondary voltage. For example, if the source voltage were 100 V, at 100-dB attenuation the voltage measured on the secondary side would be 0.001 V. At 120 dB it would be 0.0001 V. Care must be taken to ensure that the readings are not influenced by external noise, often requiring the use of a screen room to obtain accurate readings. More discussion of shielding techniques will follow later in this chapter.

Normal mode, or *transverse-mode*, as it is known in some parts of the isolation transformer industry, noise appears line-to-line at the input to the transformer. It falls into about the same category as surges described earlier. In attempting to attenuate surge voltage, the designer's goal is to design the transformer to distinguish between what is power and what is surge or noise. The key to doing this is the difference in the frequencies. The power signal is constantly low in frequency, say, 60 Hz, while surges and noise will appear at much greater frequencies. As stated earlier, and supported by the equations presented, the transformer by itself can do nothing to attenuate slowly changing voltages applied to its input, and a voltage surge that increases the line voltage by 15%, and stays there for a long period of time, will be passed right on through the transformer. But the damaging surges discussed earlier are all of the type that have fast rising leading edges, and are over with in a relatively short period of time.

Refer back to Fig. 4-1, showing the transformer equivalent circuit. In the makeup of the transformer is an element called *leakage inductance*. Just as with the coupling capacitance, there is no way to completely eliminate leakage inductance, but various construction techniques can be applied to minimize or maximize its effect. Leakage inductance comes about because not all of the flux links the two windings via the core. Figure 4-7 shows an example where flux has leaked from the iron core and completed the magnetic circuit through the air. Such leakage is associated with both the primary and secondary windings. This example has the windings on separate legs of the core, but the principle applies to any transformer, even if the windings are placed one on top of the other (layer wound), as is more usual. Placing the windings closer together to reduce the leakage inductance is achieved in some cases by interleaving the windings or splitting up the primary and secondary to get the geometric centers as close to one another as possible.

Leakage inductance resembles inductance in series with the winding of the transformer, as shown in the equivalent circuit. At power-line frequencies the effect of this series inductance is slight, but at higher frequencies the reactance

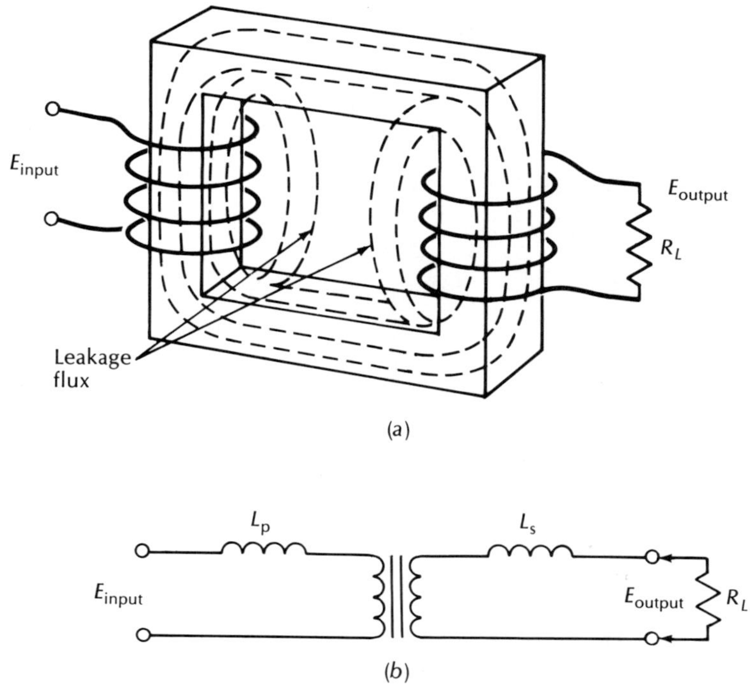

FIGURE 4-7 (*a*) Physical illustration of leakage flux; (*b*) schematic diagram showing primary L_p, and secondary L_s, leakage inductance.

is high and, being in series, will have a marked effect on the performance of the transformer. In other words, as the frequency increases on the input to the transformer, its ability to pass voltage to the secondary is diminished. The result is that a transformer intentionally designed to maximize the leakage inductance will pass line frequency voltage quite well, but any high frequency voltage riding on top of the line voltage will be attenuated, just as if it were passing through a lowpass filter. Figure 4-8 shows graphically how a typical isolation transformer, designed to maximize leakage inductance within reason, will perform. The addition of output capacitance will greatly enhance the performance, as shown.

Design Example

It is desired to provide an isolation transformer for a load. It is determined that a 500-VA transformer rating will be sufficient to run the load. Nominal input voltage is 120 V AC and the frequency is 60 Hz. Output voltage at full load is required to be 120 V AC. Input and output current will be 4.17 A, ignoring efficiency loss.

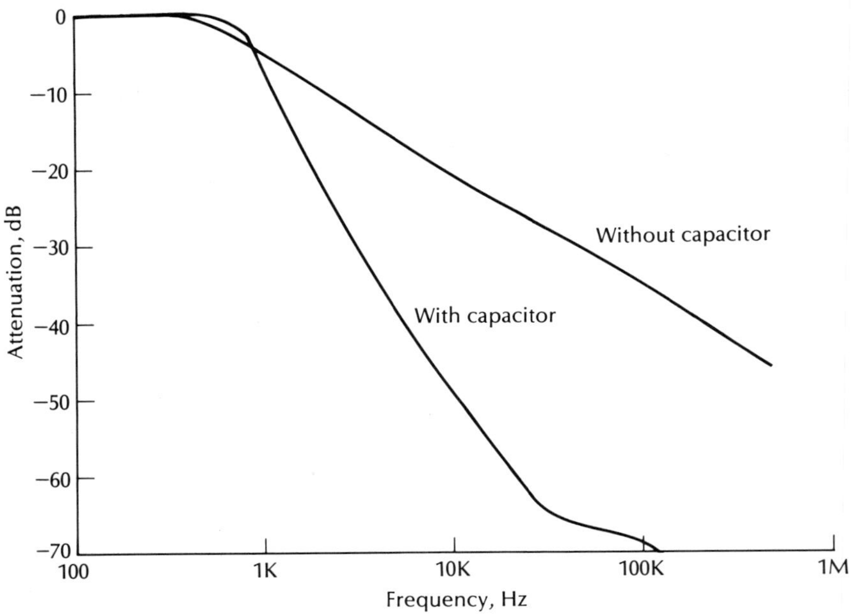

FIGURE 4-8 Transverse-mode attenuation curve.

The first task is to select a core with sufficient power-handling capability. The power-handling capability of a core takes several considerations into account. The material the core is made of will determine the maximum flux density that it will operate at, without saturating. The cross-sectional area of the core will determine how many turns must be placed on the primary to achieve the flux density desired. Finally the area of the window will determine how much copper will fit, given a need for copper wire sufficiently heavy to handle the current for the power rating of the transformer. Assume for this example that the core material is sufficient for operation of the transformer at 15,000 G. Compute the approximate core area A_c:

$$A_c = \frac{\sqrt{P_{out}}}{5.58}\sqrt{\left(\frac{60}{f}\right)} \tag{4.14}$$

where f is in hertz, P is power out in watts, and A_c is in square inches. Equation (4.14) tends to equate to slightly more core area than is actually required in a typical transformer, but keep in mind that this design will require shielding and other construction techniques that take up winding space, and the only way to provide more window space is to reduce the amount of copper, which is accomplished by increasing core area. Refer to the voltage formula given earlier, and recall the relationship between turns and core area, where all the other variables are held constant. Applying (4.7) to this design yields a requirement for approximately 4 in^2 of core. Scrappless *EI* lamination material is manufactured in several different tongue widths. The core area is a measure of the tongue width multiplied by the stack. Good design practice will keep the core geometry such that the stack is equal to or up to two times greater than the tongue width. When equal, this is known as a *square stack.* Less than square, or too much over square, will sometimes be required, as a result of package constraints where the transformer must fit, but is not generally recommended. For this design a 2-in stack of *EI*-2.0 will yield an area of 4.0 in^2. If it is determined later that less core area is acceptable, then going down to the next size core, *EI*-1.75, and adjusting the stack as required is the appropriate course of action.

Wire selection is the next step. Just as with the selection of the core, there are several decisions and considerations necessary for selection of the size of the wire for each winding. In this example the secondary voltage is the same as the primary, so we can safely assume that the same wire size will work equally well for both. The relationship between the size of the copper wire and the amount of current it must conduct is important to the outcome of the design. If the wire is too small it will fit well within the available window space provided by the core, but will cause overheating and eventually failure of the part. If it is too large, it will run cool, but will lead to a core that is larger than necessary for the design. The actual selection of the wire size depends on several considerations, such as whether the transformer will be enclosed in a box where no air flow will help cool it, or whether a fan will be available for cooling. The insulation class used is another consideration for wire size and how hot the part will be allowed to run.

Current density in the wire selected will be expressed as circular mils per ampere (CM/A). The wire's circular mil area is given in the wire table, along with other important facts. For a transformer of this power rating it is acceptable to find the wire sized for anywhere from 500 to 1200 CM/A. The wide range is due primarily to the variations in operating environment as stated above. Remembering that this particular part will be constructed with the windings boxed within shielding, which reduces the ability of the wire to cool to the surrounding environment, it is wise to start toward the higher end. A useful item to remember is that the wire's cross-sectional area will double for every three AWG (American wire gauge) size changes. A No. 14 wire will be twice the area of a No. 17 wire. For this design a No. 14 round wire is selected. This wire will provide the 1000 CM/A we're after.

The number of primary turns can now be calculated. Rearranging the voltage equation yields

$$N_p = \frac{E_{RMS} \times 10^8}{4.44 B_{max} f A} \tag{4.15}$$

As can be seen, the actual number of turns used will affect the flux density B_{max}, just as changing the core area, A_c, will affect the turns required to achieve the desired B_{max}. Keep this in mind if any adjustments to core size are made, once the turns are calculated. While the flux density is usually expressed in gauss or kilogauss, the voltage equation is conveniently used with the core area expressed in square inches; that requires the conversion of gauss to lines per square inch. Multiply 15,000 Gauss by 6.45 to obtain 96,750, and plug that number into the equation and solve for N_p. This indicates that 116 turns are required on the primary. It would seem that since the secondary is required to provide the same voltage as the input, then it would also get 116 turns. At no load this is true, but when load current passes through the windings, it causes *IR* losses to occur in both windings. The *IR* loss will cause a voltage drop that reduces the voltage to the load to something less than the no-load voltage would be, based on the turns ratio. In other words, a 1:1 turns ratio will provide a less than 1:1 voltage ratio under load conditions. The output voltage can be calculated using

$$E_{out} = \frac{N_s(E_{in} - IR_p)}{N_p} - IR_s \tag{4.16}$$

Rearranging this equation to find N_s yields

$$N_s = \frac{N_p(E_{out} + IR_s)}{E_{in} - IR_p} \tag{4.17}$$

In order to use these equations, the winding resistance must first be calculated. Wire tables will give the resistance, usually in ohms per 1000 feet, of the magnet wire selected. This resistance is typically given at 20°C, and since no transformer operates with the copper at 20°C, some adjustment must be made to obtain a truly accurate figure. Copper wire resistance will increase at the rate of 0.39 Ω/°C rise. But not yet knowing how hot the final design is going to run, it is difficult to apply this information. That leaves the choice of making a guess, or just using the figure from the wire table as a first cut, and adjusting it later if necessary. The resistance of the wire in ohms per foot can be calculated from the wire table information. But now the next problem is that you don't yet know how many feet of wire will be used. See Fig. 4-9. Given *B* as the bobbin thickness from Fig. 4-9, then

$$MTL_1 = 2(E + D + 4B) + 6.28\left(\frac{W_1}{2}\right) \tag{4.18}$$

$$MTL_2 = 2(E + D + 4B) + 6.28\left(\frac{W_1 + W_2}{2}\right) \tag{4.19}$$

$$MTL_n = 2(E + D + 4B) + 6.28\left(\frac{W_1 + W_2 + \cdots + W_n}{2}\right) \tag{4.20}$$

Figure 4-9 provides a method for calculating the mean turn length *MTL* of the windings. If the *MTL* is multiplied by the number of turns, the total amount of wire needed for the winding will be known. How full the window of the core is with wire will have an effect on the accuracy of the *MTL* calculation,

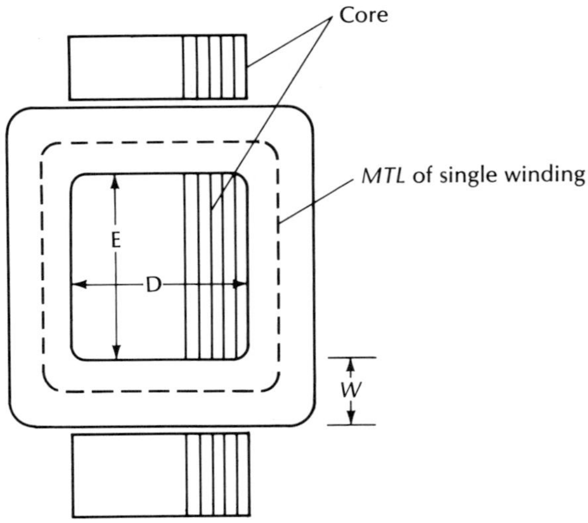

FIGURE 4-9 Diagram of single winding W on core with dimensions $E \times D$. Multiple windings are depicted in (4-18) through (4-20) as $W_1, W_2 \ldots W_n$.

but you have to start somewhere. Calculating the *MTL* for this part will yield 1.02 feet per turn. At 116 turns that means slightly more than 118 ft of wire is required for the primary winding. Using a wire resistance of 0.0025 Ω/ft calculates out to ≈ 0.29 Ω of wire resistance in the primary winding. There will be little difference in winding resistance on the secondary. Plugging in the values calculated for R_p and R_s, and the desired output voltage for E_{out}, into equation (4.17) above, N_s can now be calculated. The result can be rounded to 119 turns. Using this number in the equation for E_{out} as a check, will yield 120.6 V out at full load, for 120-V input.

It is now established that 116 turns on the primary and 119 turns on the secondary will be necessary for this design. Assuming that each winding will display about the same power loss, and thus the same temperature rise, the change of resistance due to temperature rise and its effect on the calculations just made is negligible.

The final task is to determine whether 235 turns of number 14 wire will fit within the window space allowed on the selected core. Where both windings use the same wire size it is acceptable to make a first cut using the total turns. When different wire sizes are involved, each winding must be fitted individually and then total winding space used is calculated. Part of the winding space will be taken up by the coil form, layer insulation if specified, insulation at the ends of the windings, a wrapper around the entire coil, again if specified, and in the case of this transformer shielding material. To begin with, let us just look at the fill without regard for shielding. Each layer of wire must fit within the long dimension of the window. See Fig. 4-10. The *EI*-2 core window is 1 in × 3 in long. Forty turns of No. 14 wire will take approximately 85% of the 3-in dimension, leaving adequate space for insulation. At 40 turns per layer, 235 turns will require 6 layers of wire. The total build for this part

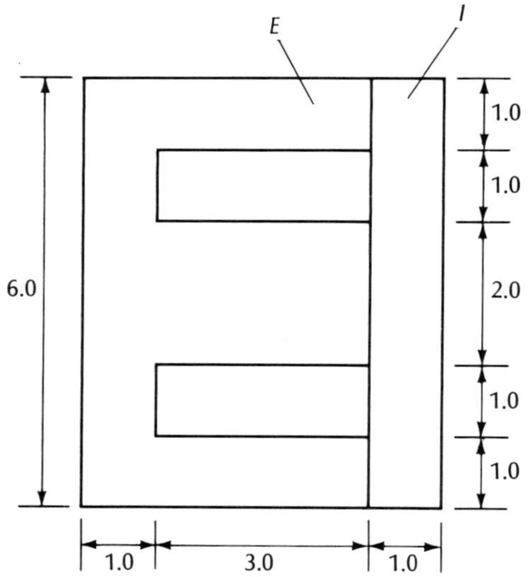

FIGURE 4-10 *EI*-2 scrappless lamination.

with 6 layers of wire, allowing for adequate layer insulation and coil form will only take up about 50% of the 1-in window dimension. This should be more than adequate for the space that will be required to properly shield the coils. As stated earlier, the core selected initially may be more than that actually required in the final design, and the steps provided above in arriving at a fill factor for the core used will have to be repeated should another core be selected. Finally, the actual construction of the coils should be studied prior to attempting to go back and make any type of adjustment to the core size.

With the construction methods presented in the following discussion we are attempting to accomplish two goals. First, there is the matter of shielding the coils to minimize the coupling capacitance between windings, thereby maximizing the common-mode noise attenuation. Second, we want to construct the transformer in such a way as to take advantage of the leakage inductance, maximizing the amount of normal or transverse mode attenuation at higher frequencies.

Figure 4-11 shows a single coil, say, the primary for the transformer from the earlier example. By dividing the two windings so each takes up half of the window, each coil can be shielded and mounted to the core individually, side by side as shown in Fig. 4-12, or in a nested configuration as shown in Fig. 4-13. The only real advantage in using the side-by-side approach is that both coils can be wound on the same-size mandrel. The shield (Fig. 4-11) is made of a strip of aluminum foil. The width of the strip is important only in how well it allows the person winding it on to do the job. Too narrow, and it takes longer to get the full coil wrapped; too wide, and it is difficult to lay in evenly, particularly around the corners. Generally a layer of mylar tape is applied to the coil first as a base for the foil. The foil is then spiral-wrapped onto the coil with another layer of mylar tape. The tape is slightly wider than the foil and allows the foil to overlap, but do so without shorting out to itself. It is impor-

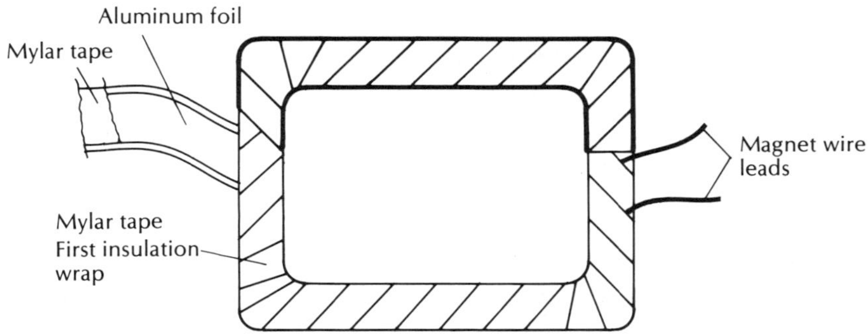

FIGURE 4-11 Foil shield added to coil of transformer.

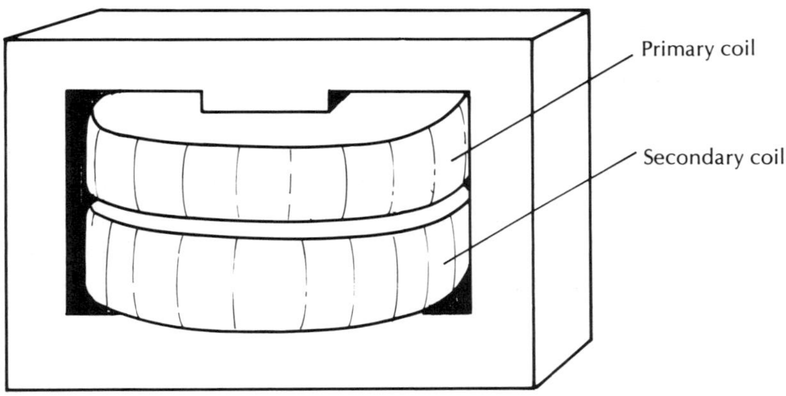

FIGURE 4-12 Side-by-side coil construction.

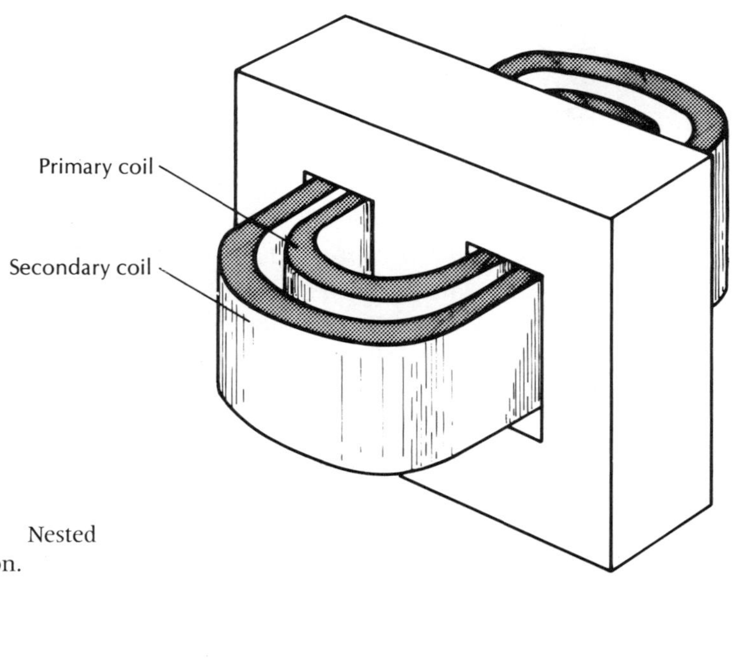

FIGURE 4-13 Nested coil construction.

tant to make certain that each turn of shielding foil is electrically isolated from the adjacent turns, or the result will be disastrous. Assume that the transformer is designed with 1 V per turn. When 120 V is applied to the primary, there will be a voltage induced into the shield as well as the secondary. That voltage will also be 1 V per turn of foil strip. If two adjacent turns are allowed to short-circuit together, then only the resistance between them limits the current that can flow between them. In this case there is little resistance so the result will be lots of current, heat, melting of mylar tape insulation, and ultimately a hole in the foil shield. Even if the problem fixes itself by burning away the short-circuit, and no more smoke appears, the integrity of the shield has now been violated and its ability to do its job greatly impaired. Once the foil has made a complete circuit around the coil, all that remains is to attach a drain wire. The best approach is to solder a wire to a small square of copper foil and then tape that to the exposed aluminum foil. Finally, finish off the coil by covering all remaining exposed aluminum foil with the mylar tape. The coil is now ready to join with another one on the core of the transformer. This method will work equally well with both the side-by-side and nested designs. This example results in a single box shield for each coil. If the drain wire for the primary is attached to the input green wire ground, and that of the secondary is attached to the load ground, the result will be a very high level of isolation. If even more isolation is desired, a Faraday shield as described earlier can be attached to either of the coils. This can be done simply by adding another layer of shield over the one already attached. This second shield will serve as the Faraday shield, and a drain wire will be attached to it just as with the others.

The shielding and construction methods described above are effective, but do prove to be costly to implement. There is a great deal of hand winding of the foil and tape, and a lot of room for error. An alternative method is to add the shielding directly as the coils are wound. This method takes advantage of the fact that the coil of the transformer is already on the winding machine; therefore, if you can manage to shield the coil at the same time as it is wound, you can save time and cost in handling. The method of winding one coil directly on top of another is called layer winding. It makes much better utilization of the window space and can result in a more efficient design than the side-by-side and nested designs described above. The disadvantage is that layer winding minimizes the leakage inductance compared with the other winding methods, so special action must be taken to enhance the leakage inductance with layer winding. Shielding is accomplished in much the same way as the previous methods. The difference here is the aluminum foil is placed in position on the mandrel before any wire is added. The foil is just wide enough that after the coil is completed it can be folded up and over the coil. Another layer of foil is added on top of the wire and folded down and over the bottom foil. Insulation is now added over the foil and another layer of foil added to start the second coil. The second coil is finished off in the same way. If care is taken in applying this method a transformer can be constructed that has very good noise isolation and is efficient to build. The problem of leakage inductance is solved in this design by the addition of strips of ferromagnetic material between windings. Some small *I* sections of lamination material can be taped together and taped to the outside of the first coil, prior to starting the second coil. This will cause a slight budge in the coil, but that is of little concern. The amount of leakage inductance achieved will depend on the amount and type of *I* material used. This method has proven to be as effective as the other types of construction against normal-mode noise.

Summary

Problems caused by the existence of EMI/RFI noise on power lines can often be very subtle and may take a great deal of time to solve. Computers, and computer-controlled machinery, are quite susceptible to noise on the power lines, and serious problems can result if not corrected. The source of the energy can be external to the building or can be generated from within. Power tools, large rotating machines, transmitters, autoignition, and so forth can all cause noise to appear on lines that feed sensitive equipment. Power-line filtering is an effective means of reducing both common-mode and normal-mode noise to levels that are harmless to sensitive equipment. This chapter has presented two ways to achieve the filtering, through the use of a dedicated RFI filter, and the use of an isolation transformer, designed in such a way that it will provide filtering as well as isolation. From a cost standpoint, the line filter is much cheaper and where true DC isolation is not important is usually the chosen method for noise reduction.

Surges are more destructive to hardware than noise, and several methods were presented for reducing surges to levels that will not harm most equipment. A standard (IEEE-587) created to describe worst-case conditions expected on indoor and outdoor power lines has provided designers with a waveform against which their equipment can be tested for damage when such a waveform is applied. This same waveform is used by designers of surge suppressing devices to determine whether the surge is being reduced, and to what level. As with power-line noise, surges can be caused by forces external to the building or can be caused by the operation of equipment within the same building. Good power management, particularly the power to sensitive or expensive equipment, will keep noise and surge problems under control, at least from the standpoint of what is generated from within the building. External forces are harder to deal with at the source, require the time and cost to fully understand what is happening, and add the appropriate noise and surge reducing devices required for full protection.

As we get further into this book the types of power-line protection described will become more complex and expensive. Additional complexity results from the need to provide more complex protection, as will be found in Part III. Power-line regulation can, and often does, include noise and surge reduction at the same time. While the actual implementation of noise and surge reduction as a part of the design of a line regulator is not discussed, it is assumed that any of the methods discussed in this part can be applied to the remainder of the book.

PART III

Power Line Regulation

The earlier chapters of this book dealt primarily with the issue of cleaning up the AC power line and preventing potentially damaging noise and impulses from getting into equipment. This part will deal with methods of regulating the power-line voltage to keep it within acceptable limits. Some of the solutions presented here will include many of the same noise and surge attenuation characteristics already presented, but, for obvious reasons, will be more expensive to implement due to the addition of line regulation capability.

The word "regulation" generally implies to most engineers that a voltage or current is being held to within a very limited range while the line and or load is busy changing over a specified range. In AC voltage regulation products, the limited range of regulation can be anything from less than 1% change, over a given line and load variation range, to 10% or greater, depending on the need and the technology implemented. In this section we will look at several methods of achieving AC voltage regulation and attempt to point out the pros and cons of each. Some design examples are presented, but by no means are to be considered the only or best method of accomplishing the task. Rather, they are presented as a means of conveying ideas and information.

CHAPTER 5
Thyristor Regulator

The thyristor is much like a normal rectifier that has been modified to block in the forward direction until a small signal is applied to the control or gate electrode. After the signal is applied, the device conducts in the forward direction with a forward characteristic very similar to that of a normal silicon rectifier and continues to conduct even after the control signal has been removed, so long as sufficient forward current, holding current, continues to flow. Thyristors, in the form of either a silicon controlled rectifier (SCR) or a triac, have been in use for the control of AC voltage for a long time. Individual characteristics of the two devices make each better suited for different roles. Because the triac is capable of conducting current in both directions, it is commonly used to control the AC voltage to appliances such as a lamp dimmer. Figure 5-1 shows such an application.

When the timing of the gate trigger circuit is changed in relation to the AC line, the triac will delay its turn on after each zero crossing. This delay will result in a reduction of the RMS voltage to the load, in this case a lightbulb, which results in reduction of the light intensity. Figure 5-2 illustrates the voltage waveform present at the load while the dimmer is "turned down," reducing the light intensity.

The SCR, on the other hand, because of its ability to conduct current in only one direction, is more commonly found in circuits where its diode-like characteristics will allow it to replace a diode, but provide control at the same time. An example of this is shown in Fig. 5-3, where two SCR devices have replaced two of the four diodes in a full-wave bridge rectifier. By controlling the gate drive to each of the two devices, the voltage out of the bridge, which usually feeds either a capacitor, an inductor, or both, can be controlled or regulated. These examples are very basic ways of regulating an AC voltage. In the one case of the lamp dimmer the AC line voltage is directly fed to the triac, while the more common practice, where the full-wave bridge is used, is to reduce the line voltage with a transformer prior to feeding it to the bridge. This is due only to the fact that more often than not the final voltage required out of the bridge is somewhat less than what can be taken directly from the power line. Other considerations such as isolation often come into the picture as well.

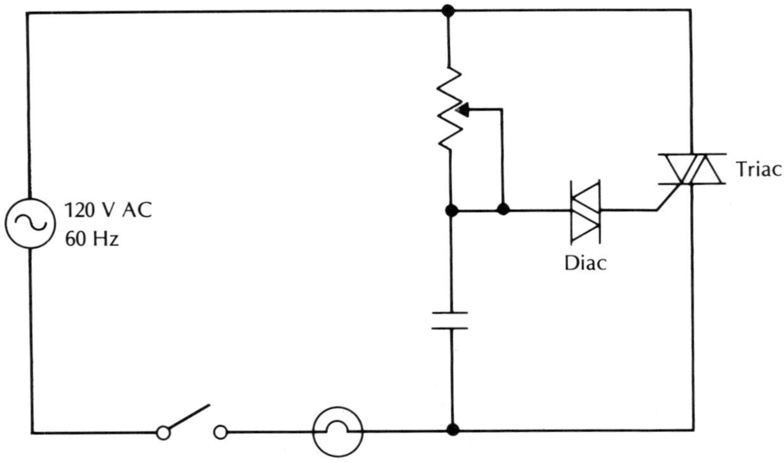

FIGURE 5-1 Basic lamp dimmer circuit using a diac to control the gate of a triac.

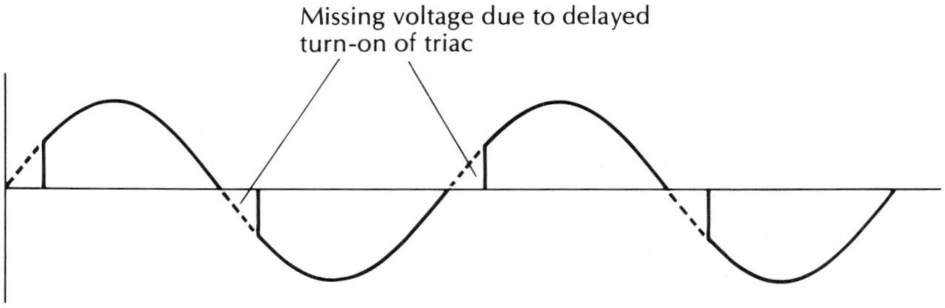

FIGURE 5-2 Voltage waveform across lamp shown in Fig. 5-1.

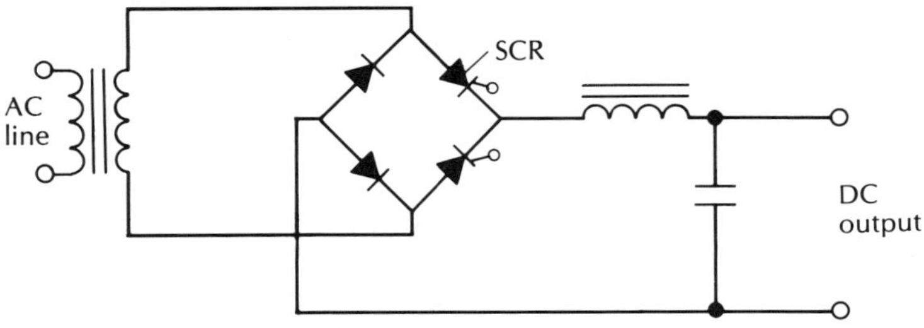

FIGURE 5-3 Typical use of SCR as control element in AC-to-DC converter. By controlling delay turn-on of the SCRs, the DC output is controlled.

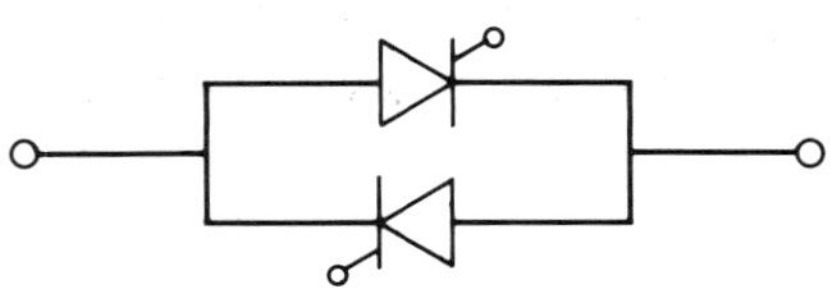

FIGURE 5-4 Two SCRs configured as bidirectional switch, similar in operation to a triac.

The triac and SCR are by no means limited to the examples presented above, and in many cases one can serve in place of the other and it often becomes a consideration of specifications of the individual parts that will dictate which is used. For example, in higher-current applications triacs disappear, since they simply aren't available at ratings as high as SCRs. But given the need to do so, a pair of SCRs arranged as depicted in Fig. 5-4 will do very nicely for providing bidirectional current flow as found in a triac. Since each has its own gate, this arrangement must be handled a little differently than the single triac gate. The topic of gate drives will be discussed in further detail later in this chapter.

The methods of regulating the AC line voltage described above give rise to the possibility of going one step further with the concept. Going back to the lamp dimmer circuit for a moment, it would be possible to replace the manually adjustable circuit with a feedback circuit that would monitor the output voltage and automatically adjust the gate drive to the triac to keep the load voltage constant. This would work for keeping the voltage from going too high, but unless it were desired to operate the load at a nominal voltage well below the 115 V AC it was designed for, there would be no way to increase the load voltage during a brownout condition. And that is one of the more critical as well as desirable characteristics of a line regulator, that is, to not only reduce the voltage during times of high line but also boost it during low line periods.

Figure 5-5 provides one method of solving this problem. By using a transformer to increase the line voltage, say, 10%, the regulator circuit will now be working all the time to lower the load voltage during times that the line is normal to high, and it will then allow more line voltage through during times of low line, effectively holding the load voltage constant over a fairly wide

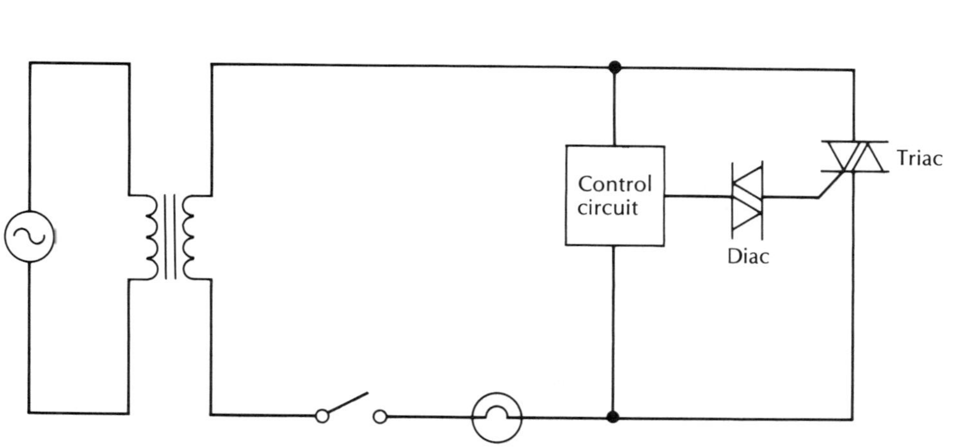

FIGURE 5-5 Lamp dimmer with control circuit and input transformer added to allow automatic regulation of lamp voltage.

range of input. The load RMS current I_{RMS} at any delay angle is given in terms of the normal full-load RMS current at zero delay angle, $I_{\mathrm{RMS}\,0}$ by

$$I_{\mathrm{RMS}} = I_{\mathrm{RMS}\,0}\left[1 - \left(\frac{\alpha}{\pi}\right) + (2\pi)^{-1}\sin 2\alpha\right]^{1/2} \tag{5.1}$$

The load RMS voltage bears the same relation to the full-load RMS load voltage, substituting E for I in (5.1). This shows that, although the theoretical delay range for complete control is 0–180°, a practical range of 20–160° gives a power control of about 99% to 1% of maximum.

The circuit in Fig. 5-6*a* requires a trigger pulse source having a pair of isolated outputs, insulated from each other by at least the peak value of the line voltage, and the two trigger signals must be 180° out of phase with each other and must shift together, with respect to the line voltage zero crossing, when the power throughput is adjusted. Provided that two separate pulse signals are used, it is permissible for the two to be identical, operating at twice

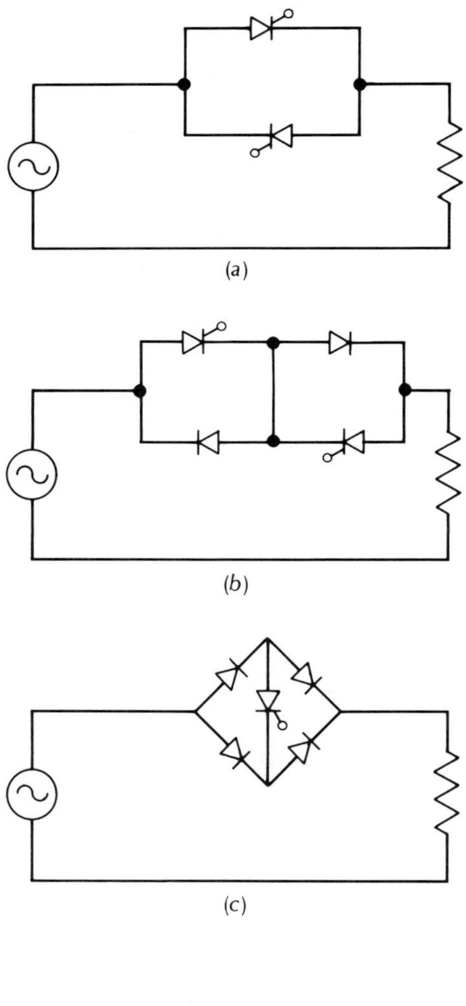

FIGURE 5-6 Three basic circuits that provide single-phase AC control.

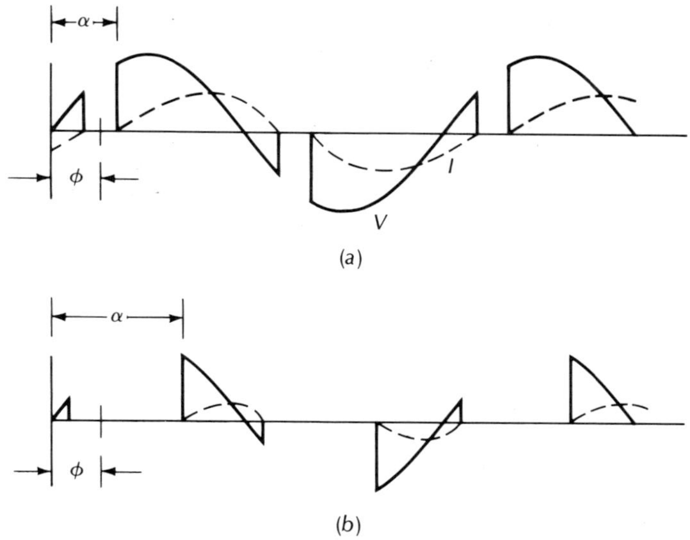

FIGURE 5-7 Voltage and current with reduced conduction, $\alpha > \varnothing$.

the supply frequency. The thyristor whose anode–cathode is forward-biased will fire when pulses are applied to both. Normally it is not considered safe to drive a thyristor gate positive while its anode is reverse-biased; however, in this circuit it is permissible since the thyristor that is fired immediately removes the reverse voltage from the other. Figure 5-6*b* is a variation of the circuit that eliminates the necessity for isolation of the two pulse sources from each other. The common thyristor cathode connection permits the use of a simple three-terminal push–pull pulse source. The advantage is gained, however, at the expense of two extra diodes, each of which must be rated for the same current and voltage as the thyristors. There will be twice the heat dissipation compared with the simpler circuit of Fig. 5-6*a*. Since no reverse voltage, or at least not more than 1 or 2 V, can be applied to the thyristors in the circuit of Fig. 5-6*b*, this circuit may be preferred if transient spike voltages are likely to be present on the power line. The only resulting breakdown in the thyristors must be in the forward direction, which is not as dangerous as reverse breakdown because the thyristor is automatically turned on by the over voltage and the voltage across it rapidly drops to zero.

Another variation is illustrated in Fig. 5-6*c*. Its main advantage is the use of a single thyristor, with the consequent simplification of trigger circuits. Its main disadvantage is that the conducting path in each direction contains three devices, two diodes and a thyristor in series, with relatively high power dissipation.

With inductive loads, as shown in Fig. 5-7, there is a sharp transient change of load voltage at the end of each current loop. This may not be important to the load, but it does stress the thyristors. Thus, at the end of the first positive loop of current, as shown by the dotted waveform in Fig. 5-7*a*, the conducting thyristor turns off, thereby disconnecting the load from the supply, resulting in the voltage at the load rapidly returning to ground. The effect of this transient voltage on the circuits of Fig. 5-6*a* and 5-6*b* is to apply a sharply rising positive anode voltage to the thyristor opposing the one which has been conducting. If the rated *dv/dt* of the device is exceeded, the thyristor will then turn on and conduction will take place independent of the gate drive. The rate of rise of voltage may be controlled by adding a resistor in shunt with the load. The rate of rise of voltage will then be determined by the time constant L/R, where L is the load inductance and R is the series sum of the load resistance and the added resistor. This resistor has the further advantage of providing a direct resistive path for thyristor current so that the holding current is established quickly after turn-on, instead of at a time determined by the rate of rise of load current, which is fixed by the applied voltage and load inductance.

Whenever a transformer is placed between the thyristors and the load it is important to ensure that the two thyristors are fired exactly at 180° relative to each other. If this is not the case, the positive and negative current loops will differ in magnitude and a resultant DC will flow through the primary of the transformer. For this reason it is unsatisfactory to determine the firing phases by independent trigger circuits using, for example, separate trigger diodes, which may differ in breakdown voltage.

While this is a fairly cost-effective method for regulating AC voltage, some loads may be sensitive to the resulting waveform shown in Fig. 5-2. The result of delaying the turn-on of the thyristor causes a chunk of the AC sine wave to be eliminated. While controlling the RMS voltage this also causes a great deal of distortion to appear at the load. The lamp obviously is not overly concerned with this, but some electronic devices could be, and it is important to keep this in mind when considering this as a method of providing regulation to a particular load.

CHAPTER 6

Constant-Voltage Transformer

For a great part of this century engineers have been using the nonlinear characteristics of magnetic materials in their designs, allowing them to perform tasks otherwise best suited to circuits containing active power devices. The magnetic amplifier, for example, uses a saturable transformer, or reactor, in a circuit to amplify power. Nonlinear magnetic circuits are useful for several functions, including, but not limited to, waveform conversion and voltage stabilization, the characteristic we are most concerned with here.

In principle the concept of a constant voltage transformer is quite simple. Figure 6-1 shows an example of a simple ferroresonant regulator utilizing two-coil construction. A linear inductor L and a capacitor C are placed in series across the power line. The primary voltage of the transformer, and thus the voltage across the capacitor, is allowed to increase during each half-cycle until a predetermined amplitude is reached. At that point the transformer will saturate, and its normally high impedance characteristic suddenly becomes a low-impedance path, preventing further voltage increase across the capacitor. The result is a fairly flat topped waveform as illustrated in Fig. 6-2.

The turns ratio of the transformer can be set to provide the desired output voltage to the load. The inductance is provided by designing the primary winding and core relationship as you would a gapped inductor. There are many variations on this basic theme, some getting quite complex in the use of multiple windings, shunts to enhance the effects of the inductance, and stepup windings to reduce the size of the capacitor for a given design. Figure 6-3 illustrates this by modifying the circuit of Fig. 6-1 to add additional turns to the primary winding, thus providing more voltage to the capacitor, reducing the capacitance requirement.

For most general-purpose applications the constant-voltage transformer will do quite well. Proper design can result in good load and line regulation, and fairly quiet operation, a common complaint from users of this type of regulator. Modern designs will provide up to 90% efficiency, provide good response to nonlinear loads, and provide fairly good regulation, in the range of 3–5%, for an input line change of +15 to −20%. Total harmonic distortion of the output waveform can be held to less than 5%, with no single harmonic exceeding 3%. Additionally, proper construction techniques to pro-

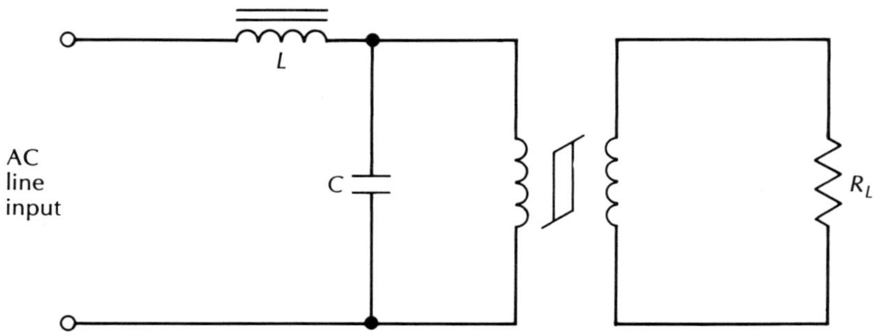

FIGURE 6-1 Basic ferroresonant AC line regulator.

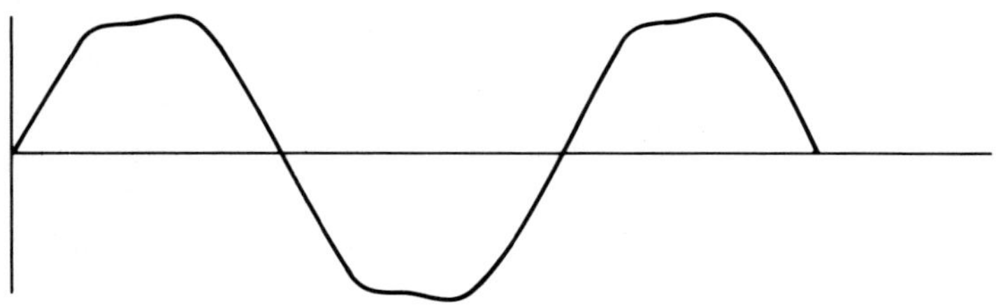

FIGURE 6-2 Voltage waveform typical for ferroresonant regulator.

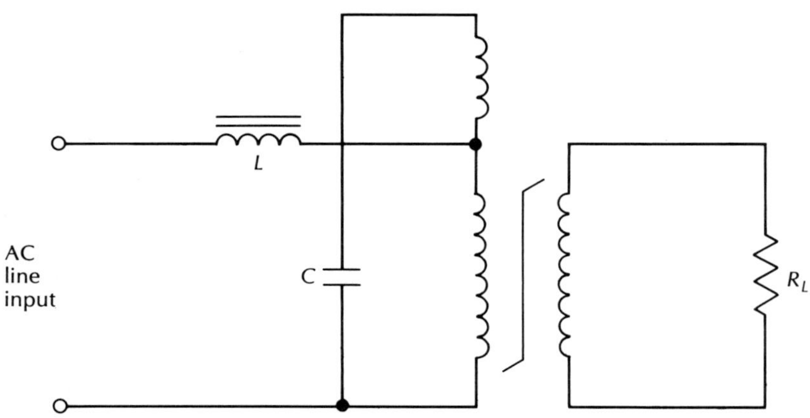

FIGURE 6-3 Ferroresonant regulator with stepup voltage.

vide shielding and high isolation will allow common-mode and transverse-mode noise attenuation of the same order as discussed previously with respect to isolation transformers. Care should be taken when applying a ferroresonant transformer in an application where the power source is a generator. Because of the nature of the operation of the transformer, utilizing its saturation characteristics to regulate the output voltage, any fluctuation of the input frequency will result in a change of output voltage. Figure 6-4 describes the

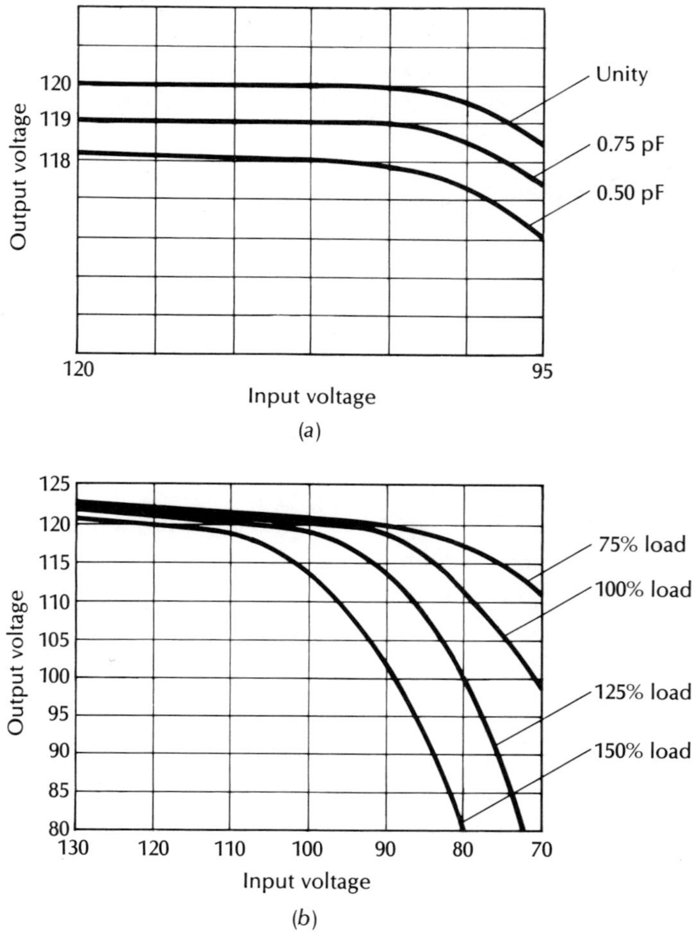

FIGURE 6-4 (*a*) Output-voltage versus power factor load; (*b*) output voltage versus percent load.

operating characteristics typical for a modern ferroresonant line regulator for output voltage versus load power factor and output voltage versus percent load. These curves are not intended to describe the technology in total capability, but rather are presented to show the typical performance of a particular design. As mentioned earlier, construction plays a significant role in determining the output characteristics of a constant voltage transformer.

CHAPTER 7

Tap Changer

If a transformer were designed to have a 115-V primary and, to make things easy, were wound with 115 turns, the secondary voltage would then be determined by the number of turns, ignoring losses for the moment, as 1 V per turn. A secondary winding then with 130 turns would provide 130 V to the load, when 115 V is applied to the primary. Now, if the secondary were constructed so that every turn were brought out as a tap that the load could be wired to, you would have considerable flexibility in selecting how much voltage would be placed on the load for a given input voltage. If only half the winding were tapped at each turn, you could select a voltage of between 60- and 130-V output, for 115-V input. Figure 7-1 illustrates such a transformer, with four secondary taps connected to a switch. The switch will allow manual selection of voltage to apply to the load. This is a very basic form of a tap changer design. As the input voltage from the power line may vary over a wide range, selection of various taps will allow the load voltage to be regulated, within the limits of the number of turns between the taps. In the case where 60 of the 130 turns have a tap, and the transformer is constructed to provide 1 V per turn, the output or load voltage can be regulated in 1-V increments. In the case where there are only four taps, as in Fig. 7-1, far less resolution is realized.

As discussed earlier in this section, the term regulation can mean a lot of things to a lot of people, and in the case of a tap changing regulator the end result will be determined by what you are willing to spend. Both the trans-

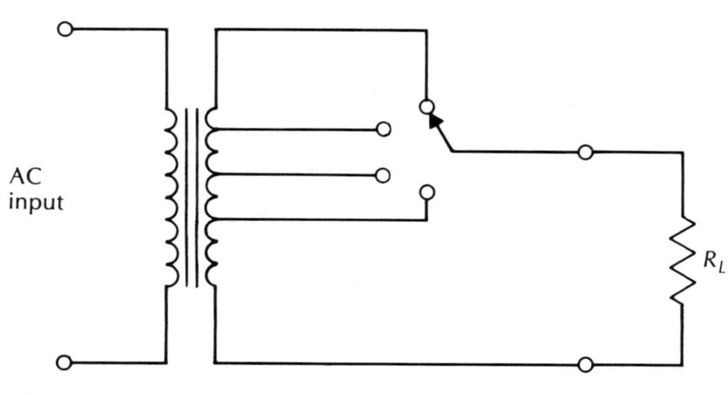

FIGURE 7-1 Manual tap selection.

former design and the method of making the actual tap selection play a major role in the total cost of a tap changer, as they are the major elements of the design. Therefore, it is critical to study the application for its actual regulation requirement before setting out to specify or design a tap changer.

Transformer Design

Autotransformer

The selection of the transformer for a tap changer design is critical from several positions. If simple regulation is all that is required, with no concern for noise attenuation or isolation, then an autotransformer is the best choice. Figure 7-2 shows schematically how an autoconnected transformer provides stepup or stepdown of input voltage. An autotransformer demonstrates several advantages over an isolation transformer designed to perform the same task. Of these, perhaps the most important is the reduction in size for the autotransformer for the same task as the isolation transformer. As compared with the isolation transformer, the autotransformer allows greater input or output current, because the winding that is common to both output and input carries a current that is equal to their difference. The overall voltage is greater than that of either the common winding or the extension arm, as shown in Fig. 7-2.

The improvement in performance over the isolation transformer is shown by the expression

$$\frac{va_{\mathrm{L,auto}}}{va_{\mathrm{T,iso}}} = \frac{\text{total voltage}}{\text{extension} - \text{arm voltage}} = \frac{E_{\mathrm{h}}}{E_{\mathrm{h}} - E_{x}} \tag{7.1}$$

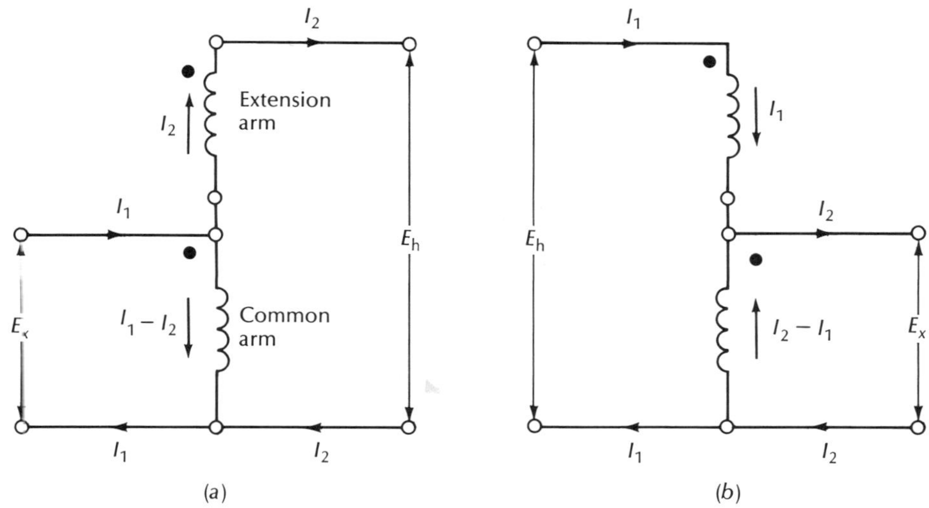

FIGURE 7-2 Autotransformer: (*a*) stepup; (*b*) stepdown.

where E_h is the higher voltage and E_x is the lower voltage. In the stepup transformer of Fig. 7-2*a*, E_h is the secondary voltage, while in the stepdown transformer of Fig. 7-2*b*, it is the primary voltage. By substituting

$$n = \frac{E_h}{E_x} = \frac{1}{r} \tag{7.2}$$

into (7.1), the relationship

$$\frac{va_{L,auto}}{va_{T,iso}} = \frac{n}{n-1} = \frac{1}{1-r} \tag{7.3}$$

shows the improvement is the greatest when the difference between input and output voltage is small. The rating of a 100-V-A isolation transformer is increased to 200 V A by using the autotransformer connection, if $r = 0.5$, and further to 1000 V A if $r = 0.9$.

$$\frac{\text{reg}_{\text{auto}}}{\text{reg}_{\text{iso}}} = 1 - r \tag{7.4}$$

As shown by (7.4), regulation also improves with the autotransformer. Generally what (7.4) states is that with an autotransformer the core area can be smaller, by a factor $1-r$, than that of the isolation transformer core. In a tap-changer regulator there is seldom much need to step the voltage up or down too far, so this relationship can be utilized quite well to make a very compact transformer for relatively high power regulator.

Isolation Transformer

Referring back to Part II, the isolation transformer design techniques describing shielding to attain high noise attenuation can be applied to an isolation

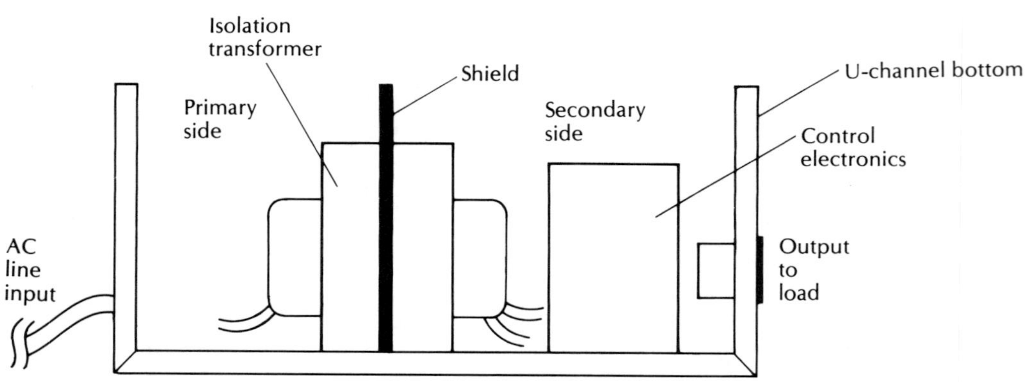

FIGURE 7-3 Simple enclosure illustrating isolation of primary from secondary components by installing a shield around the transformer core. The shield must fit tight from wall to wall and ceiling to floor to provide isolation.

transformer intended for use in a tap-changer regulator circuit. Care must be taken to maintain full input to output isolation beyond just the transformer. Since other circuit elements will now be involved, it is important to keep all primary circuitry isolated and shielded from secondary circuitry.

Figure 7-3 is an example of how an enclosure should be designed to maintain the high input-to-output isolation needed. By counting on the internal shielding of the transformer itself to attenuate noise through the transformer, the only remaining task is to shield the primary components from the secondary components.

CHAPTER 8

Power and Control Elements

As described earlier for illustration, it would be possible to connect the taps from the output of a transformer to a switch and then manually select the desired output voltage. If, instead, each switch position were replaced by a semiconductor device that could be switched on and off, when the tap it were connected to was the desired source, then a fully electronic control method could replace the manual switch.

The devices most commonly devices used for switching each tap on and off are triacs and SCRs. These electronic devices work well for this application, and if properly applied can give reliable performance. Figure 8-1 shows a simple two-tap approach using triacs for tap selection. A common difficulty with using triacs in this type of application is the comparatively low *dv/dt* rating of the device. The device is rated in either of two ways. (1) the *dv/dt* rating the device will withstand when applied with the gate signal off, and no current flow, or (2) the *dv/dt* rating the device will withstand when applied with the gate signal off, with the device attempting to block current that has been flowing, such as what happens at a current zero crossing on the power line. Since the second of these two conditions happens every half-cycle for the device that is running the load, it is of importance when it is desired to stop current flow from one switch at the end of the half cycle, and begin with another, without any interruption of load current. If the switch that has been conducting should fail to turn off as predicted at the zero crossing, there will be a half-cycle of high current flow in the two devices as a result of effectively short circuiting the turns of the transformer. This high current flow could damage the parts, and will possibly have an effect on the operation of the load.

An SCR with the same current and voltage rating as a triac will typically have a higher *dv/dt* rating. Particularly during the critical period when it is time for one switch to turn off and connect another tap to the load, the SCR will be superior. By connecting two SCRs as shown in Fig. 8-2, they will replace one triac. Since one SCR has to conduct current only one way, neither has to interrupt current flow and suddenly block, as is the case with the triac. For this reason the SCR switch will work much more reliably in the application.

One drawback to this approach, aside from the additional cost of two parts to make one switch element, is that there are now two gates that must be driven. This can be accomplished several ways. Figure 8-3 illustrates two methods of accomplishing isolated drive to two inverse parallel SCRs. Both methods provide isolated drive that will turn on the devices by a signal from the control circuit. The Fig. 8-3*a* method uses a high frequency oscillator driving the primary of a small gate drive transformer with two isolated secondaries. The frequency can be made very high to minimize the size of the gate transformer. The secondary voltages on the transformers need be only a couple of volts to turn on the part. Current requirement will depend on the part itself and can be found in the data sheet for the SCR. Whenever the control circuit starts the oscillator that drives a pair of gates, both SCRs will be gated on. But since only one of the two will be forward-biased at the time, only one will conduct current. If that same pair is still required to be on at the end of the half-cycle, the oscillator will continue to run and the current will reverse and flow through the other device.

The method shown in Fig. 8-3*b* uses an optocoupler to provide the isolation from the control circuit. This circuit does not use any power from the control circuit, other than that required to drive the optocoupler, to drive the gates.

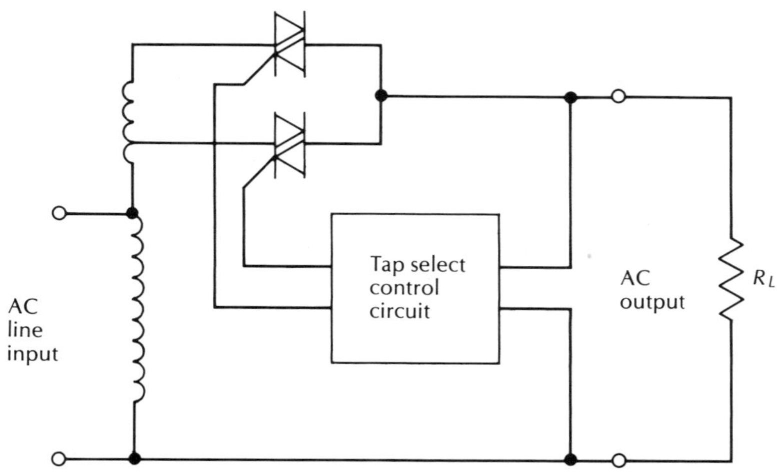

FIGURE 8-1 Two-tap regulator utilizing triac tap switches.

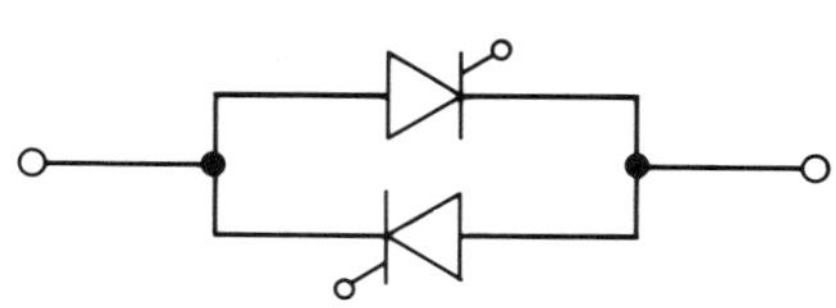

FIGURE 8-2 Two SCR's arranged as bidirectional switch to replace single triac.

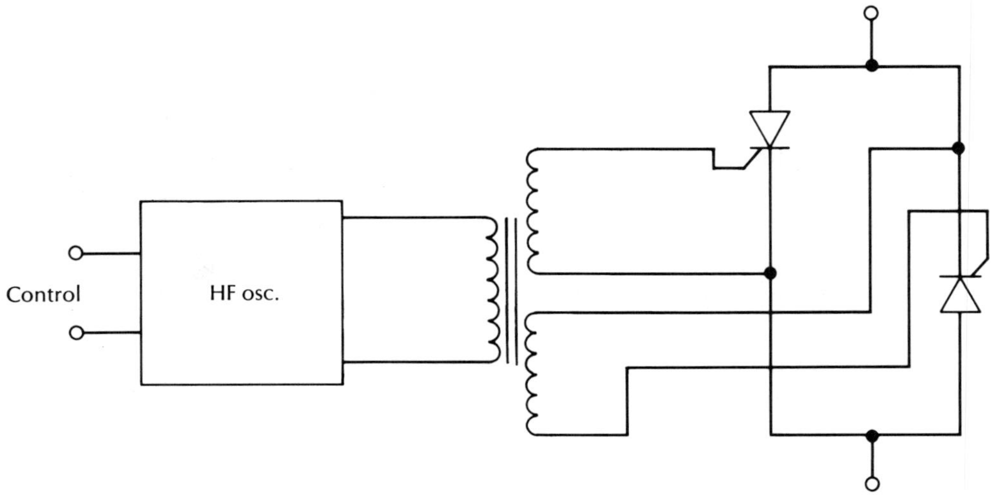

(a)

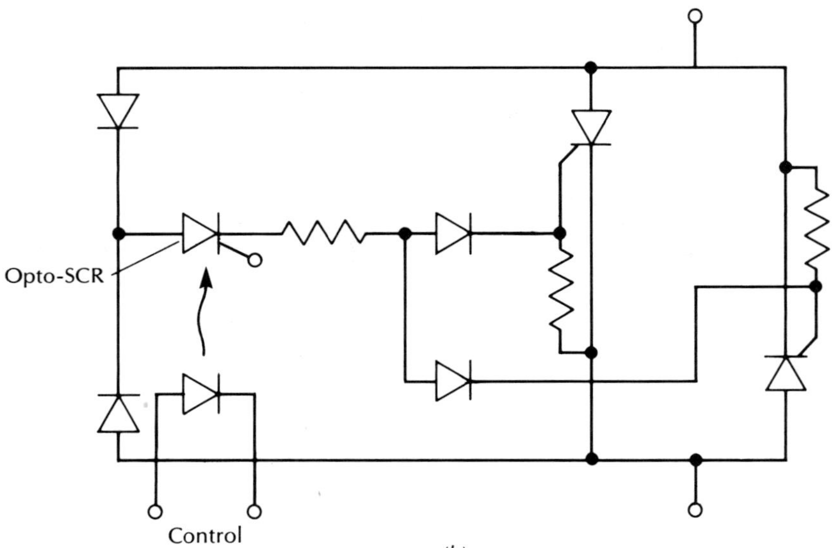

(b)

FIGURE 8-3 (*a*) High-frequency oscillator (HF osc.) provides isolated drive to SCR gate; (*b*) opto-SCR provides isolated SCR gate drive.

Instead, whenever the optocoupler is conducting it will provide gate current to the device that is forward-biased at the time, and all the required gate current will flow from the source. As soon as the SCR turns on, the voltage across the SCR is reduced to its forward drop, 1 V or so. This prevents getting full line voltage on the gate of the SCR, which will damage it.

The control circuit for the tap changer must make a decision fairly quickly as to when it is time to change taps to correct the voltage due to line or load fluctuations. It is best to monitor the output of the regulator since the actual load changes are hard to predict and better precision in the switch points can be achieved. The type of voltage measurement circuit depends on how quickly correction is needed, and how much accuracy must be maintained. A simple approach is to use a peak voltage detector to monitor the output sine wave, and compare the voltage to fixed references. Such a circuit is shown in Fig. 8-4.

While this is a simple approach, it does have the drawback that when the load is nonlinear, as is the case with switching supplies in a computer, the peak of the voltage waveform will flat-top. This will present a degree of error in the voltage detection circuit and cause the switching points to be less predictable. The combined impedance of the transformer winding, and the switch elements, will dictate how much flat topping will occur. It now becomes a matter of specsmanship and desire as to whether this method is satisfactory. The amount of error caused by the error in the waveform may result in a few percent of error in the output voltage at which a tap change takes place. In other words, if a tap change due to a reduction in input line were supposed to occur when the output reduces to 110 V, placing a higher tap on the load, thus a higher voltage, this tap change might occur at 112 or 113 V instead. Again, it will depend on the amount of flat topping on the voltage

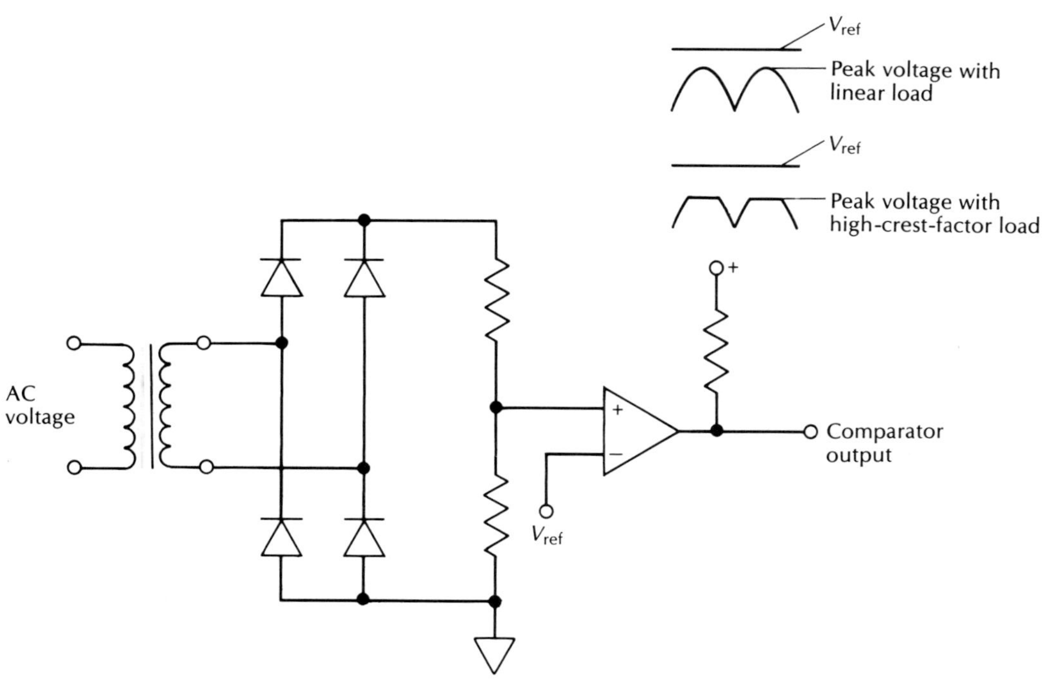

FIGURE 8-4 Peak detector–comparator output will switch high when AC voltage exceeds reference. Waveform distortion will create error in switch point.

waveform, and that depends on the load itself and characteristics of the transformer.

A more dependable, as well as predictable, approach is to use a true RMS voltage detector rather than a peak detector. Now, no matter what the effect of load on the output waveform, the control circuit will cause the tap change to occur at the same predictable point each time. Measuring the RMS value of the output can be simple if speed is not a factor. There are several true RMS to DC voltage converter integrated circuits (ICs) available that are low in cost and perform well for this type of circuit. Figure 8-5 shows the control circuit now with RMS voltage measurement instead of peak. This method is reliable, but will not provide response down to the half-cycle specified by some manufacturers of these types of products. Again, it becomes a question of what is required for the load. An alternative to the RMS to DC converter IC is to use discrete circuit components and build your own. The circuit shown in Fig. 8-6 will provide true RMS measurement and half-cycle response. The circuit works by integrating the output waveform over a half-cycle period, beginning from the zero crossing. At the same time that the sine wave is integrated a fixed DC reference is applied to another integrator. Just before the end of the half-cycle the two integrated waveforms are compared to determine whether a tap change is necessary. A system clock, or pulse generator, is provided by U_1. The output of U_1 is a square wave, in phase with the power line, and it is used to reset the integrators. Component U_4 is arranged as an integrator whose input is the AC voltage signal. Its output is expressed as

$$-K\int_0^t f(t)\, dt \tag{8.1}$$

where $f(t)$ is the AC voltage signal and K is a constant set by the input attenuator. Component U_2 is a switch used to reset the integrator. It sets the limits

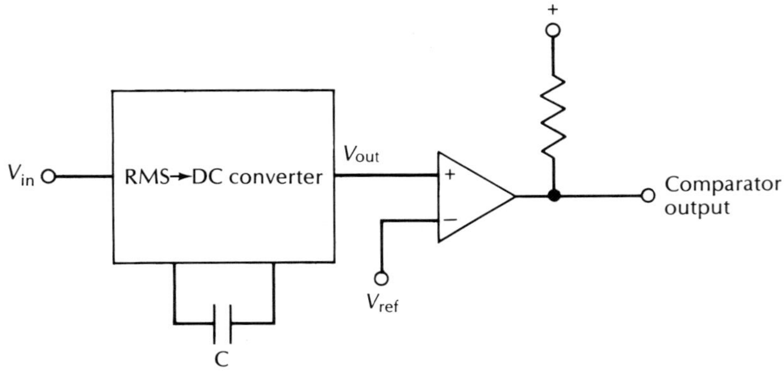

FIGURE 8-5 True RMS-to-DC converter integrated circuit provides accurate DC output regardless of waveform distortion, at CREST factors of 1–3. Response time is slowed by addition of capacitor C.

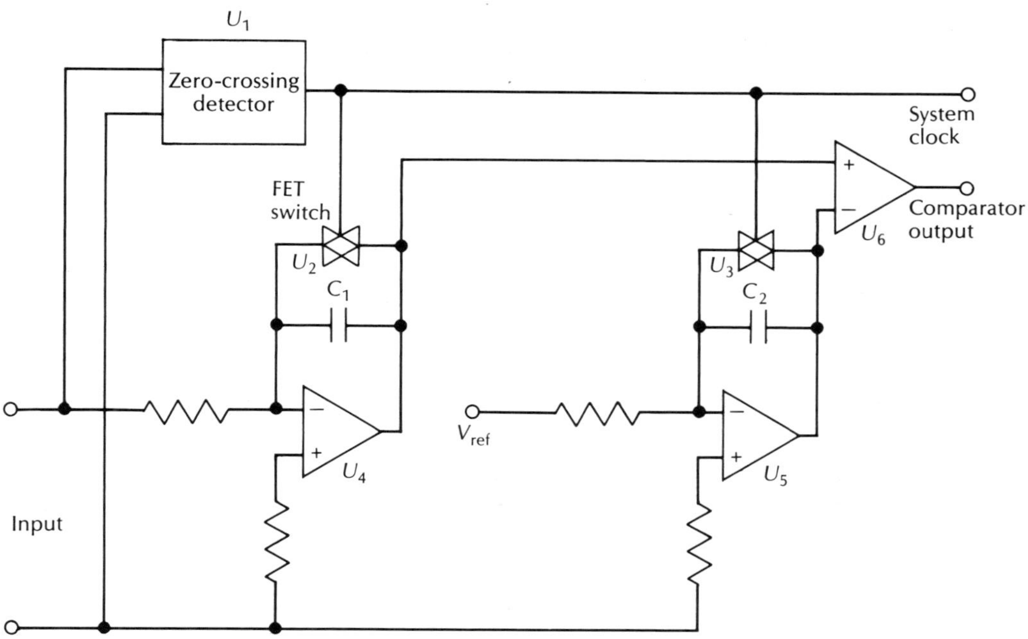

FIGURE 8-6 Fast-responding true RMS detector.

of integration by placing a short circuit across C_1 whenever the clock signal supplied by U_1 goes high. When the clock signal from U_1 is low, integration will take place. The integrated waveform corresponds to the negative half-cycle of the AC signal and stops when the signal crosses through zero and starts a positive half-cycle. At the same instant, U_1 output goes high and prevents integration until the start of the next negative half-cycle.

Component U_5 is another integrator and works similarly to U_4 as described above. But while U_4 is integrating an AC signal, U_5 integrates a DC reference voltage. Component U_3 operates from the same clock signal as U_2, and so both integrators have the same limits of integration. A comparator, U_6, is constantly monitoring the output of both integrators. As long as the output of the reference integrator is higher than the output of the AC signal integrator, the output of U_6 will be low. If this condition corresponds to the next clock signal, no change in the state of the output drive control circuit will occur. But if the reverse is true and the output of U_6 is high when a clock occurs, then the output drive circuit will take the appropriate action to change taps on the transformer to get the AC signal back within limits that will keep U_6 output low. By duplicating the AC integrator circuit, but setting it to operate over a different input range, a window detector can be made. The goal of this circuit is to keep the DC reference integrator signal at a level that falls between the upper and lower AC window integrator signals. If either AC integrator signal should cross the DC reference, a tap change will occur; which tap is dependent on whether more or less output voltage is needed, and that, in turn, depends on which integrator signal crossed the DC reference. This same arrangement can work for three-phase tap changers as well simply by arranging the clocks for three separate integrator circuits to operate in phase with their respective phase voltage to be controlled.

One problem encountered by tap changing regulators that utilize only voltage zero crossing to change to the next tap is the poor ability to deal with power factor loads, such as an inductive load. Figure 8-7 illustrates an AC voltage and current waveform. Note that the load current is lagging behind the voltage waveform. This indicates that although the AC voltage waveform has reversed at the voltage zero crossing, there is still positive current flow in the load for some additional time. If the control circuitry were to make a decision to change to another tap, and the new tap were to turn on at the voltage zero crossing point, two things would happen: (1) if the tap switches are SCRs or triacs, current will continue to flow in the first tap switch until current zero crossing occurs; and (2) the new tap switch will now attempt to pick up the load current, and for a brief period both tap switches will be on and attempting to provide load current. The result of this will be very high circulating currents in the two tap switches, as well as waveform distortion during the simultaneous conduction period.

The solution to this problem is simple. Decisions to change taps can still be made as described earlier, but hold off on turning on the next tap switch until a load current detector indicates that the current has passed through zero. The result will be very smooth transitions from one tap to the next because even though the gate signal to the first tap is off, it is continuing to conduct, and the gate signal to the next tap is held off until the current stops. There are two methods for detecting the current zero crossing. One way is to place a current transformer or some other device in series with the load. The voltage developed from this output sensor can be used to trigger a circuit to indicate when current zero crossing has occurred. An alternative is to monitor the voltage directly across each of the tap switches. This is slightly more complex than using one output detection device, but there is an advantage. Each tap switch, whether made of SCRs or triacs, will display a forward voltage drop

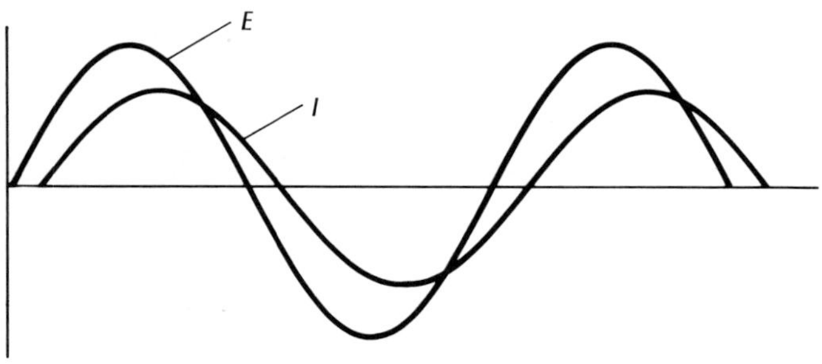

FIGURE 8-7 Voltage and current waveforms typical for an inductive load.

whenever current is flowing through the device. When a tap switch turns off at a current zero crossing, the subsequent increase in voltage across the tap switch is used to tell the control circuit that it is time to turn on the next tap. In the event that the tap that was just supposed to be turned off failed to actually block at the start of the next half-cycle, and continued to conduct load current, the voltage signal would not appear. This, in turn, would prevent the circuit from allowing any other tap to turn on. If a tap switch device had failed and short circuited, this would cause the regulator to operate only on that one tap, and, of course, it would fail to regulate the line properly, but the failure of the one tap switch would likely be the extent of the damage to the entire circuit. If two taps were allowed to be on at the same time, the failure of a tap switch could lead to even more expensive failures of other tap switches or control circuitry.

Tap switching regulators do offer a means of adjusting for line voltage variations and maintaining a relatively narrow range of load voltage while the line voltage may vary over a wide range. The examples shown up to now have been, for the purpose of simplicity, only a two-tap arrangement. Generally, the use of four or five taps will accomplish a very good regulation range. By adjusting the taps to be five to seven percent apart, a very good, cost-effective regulator can be achieved.

CHAPTER 9
Linear and Switching Power-Line Regulators

At certain times the degree of regulation provided by a tap changer as described above may not be adequate. It is often desired to have a regulator that will provide a clean, tightly regulated sine wave to the load over a very wide input range. The previous regulators described worked by manipulating the existing voltage or utilized the turns ratio of a transformer to step the voltage up or down, as required. A more precise approach is to generate the voltage waveform desired and replace the power line with that. Now the power line will simply provide the power to generate DC supplies that will, in turn, be converted into the sine-wave output of the regulator.

One approach to accomplishing this is to use a linear power amplifier to create the sine-wave output. High-power linear power amplifiers for use in AC power protection and generation equipment are rapidly being displaced by the advancement of switching technology. There are, however, some advantages in using linear as opposed to switching techniques for sine-wave power generation, so it is important to discuss the linear amplifier. A typical linear AC power source is shown in Fig. 9-1. As shown in the block diagram of Fig. 9-1, the AC power line simply provides a means for generating a positive and negative DC bus to run the power amplifier. While the technology used here is not new and efficiency is quite poor, it does deserve mention. More modern designs have replaced the linear power amplifier with a switching circuit that greatly improves efficiency, reducing size and weight. Advantages to this technology are very low harmonic distortion in the output waveform, little concern for radiated or conducted emissions, and quiet operation.

Power amplifiers are classified according to the portion of the sine-wave cycle during which load current flows. To achieve low distortion amplification, at power line frequencies, only class A, as shown in Fig. 9-2, would seem to apply. With the proper application, however, class B and class AB amplifiers can also yield essentially linear amplification. While the class AB amplifier is the design of choice for this application, we will begin with a brief discussion of the class A amplifier for purposes of comparison.

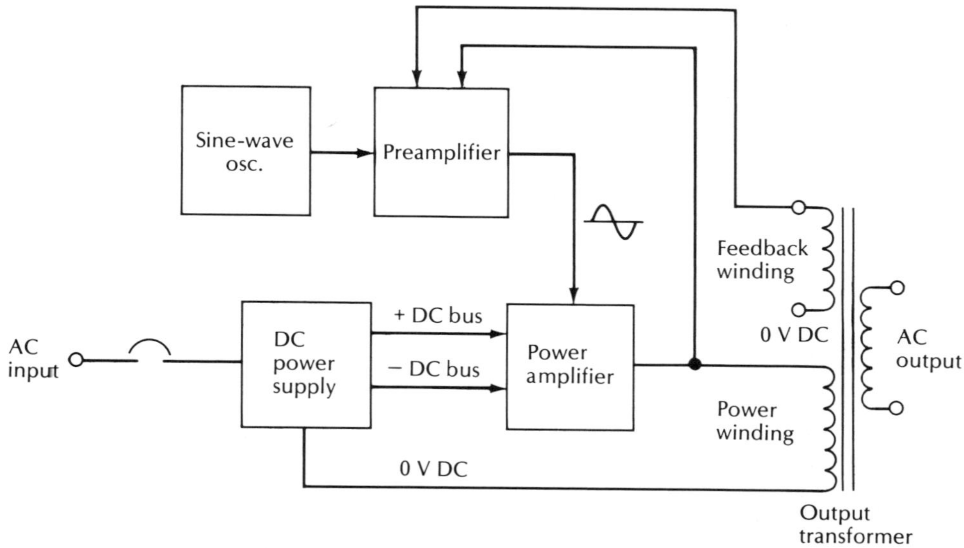

FIGURE 9-1 Typical linear power amplifier block diagram.

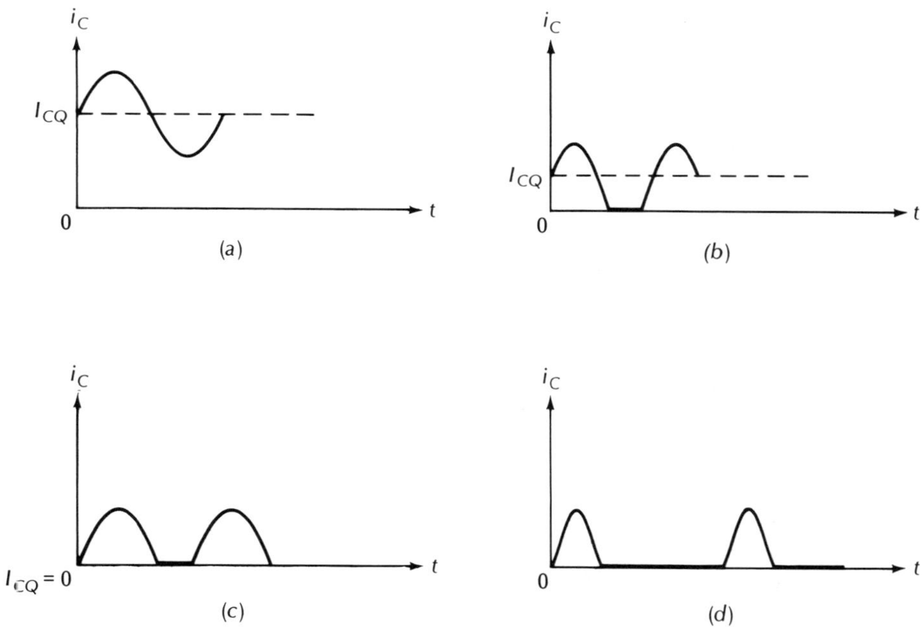

FIGURE 9-2 (*a*) Class A—current flows for 360°, full-cycle; (*b*) class AB—current flows for more than one half-cycle, but less than full-cycle; (*c*) class B—current flows from one half-cycle; (*d*) class C—current flows for less than one half-cycle.

Class A Common-Emitter Power Amplifier

A class A circuit with choke coupling is shown in Fig. 9-3. It is designed such that all capacitors are essentially short circuits and the inductor is an open circuit at power-line frequencies. At DC the capacitors are, of course, open circuits, while the inductor is a short circuit. For simplicity the inductor is assumed to have no internal resistance. Kirchhoff's voltage law is applied around the collector circuit, including only DC voltage drops, to determine the Q point. The DC load line equation for this amplifier is then

$$V_{CC} = v_{CE} + i_C R_E \tag{9.1}$$

In order to minimize bias circuit power loss while maintaining adequate Q-point stability, the emitter resistor is kept as small as possible. This results in an almost vertical DC load line, as shown in Fig. 9-4. Again, applying Kirchhoff's voltage law around the collector circuit, this time including only the AC voltage drops, yields the AC load line equation:

$$v_{CE} = -i_C R_L = i_L R_L \tag{9.2}$$

which can be written

$$i_C - I_{CQ} = -\frac{1}{R_L}(v_{CE} - V_{CEQ}) \tag{9.3}$$

Then to place the Q point for maximum symmetrical swing we use

$$I_{CQ} = \frac{V_{CC}}{R_{AC} + R_{DC}} \tag{9.4}$$

For this circuit $R_{AC} = R_L$ and $R_{DC} = R_E$, so that:

$$I_{CQ} = \frac{V_{CC}}{R_L + R_E} = \frac{V_{CC}/R_L}{1 + R_E/R_L} \tag{9.5}$$

The collector-emitter voltage at the Q point is found by setting $v_{CE} = V_{CEQ}$ and $i_C = I_{CQ}$ in (9.1) and using the value of I_{CQ} given in (9.5). This results in

$$V_{CEQ} = V_{CC}\frac{R_L}{R_L + R_E} = \frac{V_{CC}}{1 + R_E/R_L} \tag{9.6}$$

If $R_L >> R_E$, then (9.5) and (9.6) will reduce to

$$I_{CQ} \approx \frac{V_{CC}}{R_L}\left(1 - \frac{R_E}{R_L}\right) \approx \frac{V_{CC}}{R_L} \tag{9.7}$$

and

$$V_{CEQ} \approx V_{CC}\left(1 - \frac{R_E}{R_L}\right) \approx V_{CC} \tag{9.8}$$

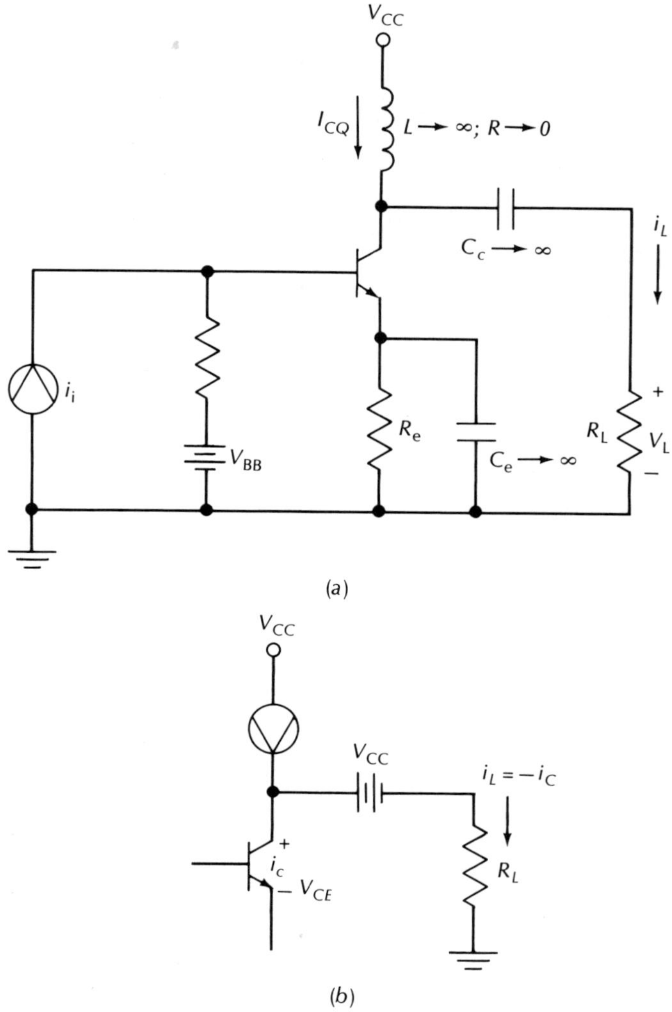

FIGURE 9-3 (*a*) Inductor-coupled power amplifier; (*b*) equivalent circuit.

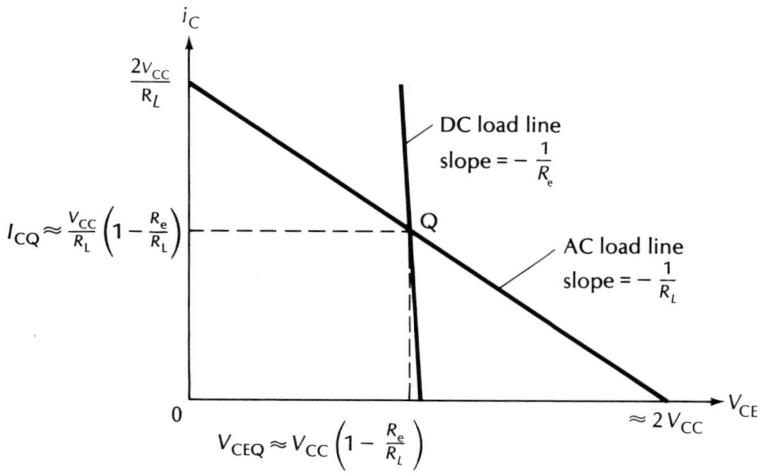

FIGURE 9-4 Power amplifier load lines.

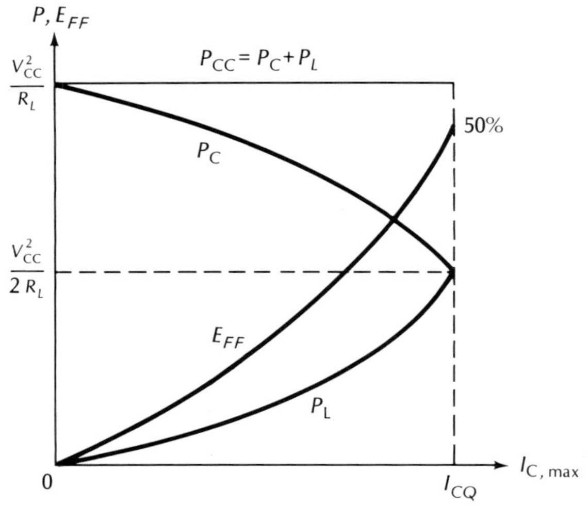

FIGURE 9-5 Variation of power and efficiency with collector current.

The AC load line passes through the Q point with slope $-1/R_L$, as shown in Fig. 9-4. Note that the maximum collector current swing is from 0 to $2I_{CQ}$ as v_{CE} swings from $2V_{CC}$ to 0, and that v_{CE} is limited by the saturation voltage of the transistor to a minimum voltage of $V_{CE,sat}$. To simplify calculations, the saturation voltage is assumed to be zero. It is interesting to consider how the collector-emitter voltage can become twice the supply voltage. Since the inductance is very large, no AC current will flow through it, and for purposes of analysis, it can be replaced by a constant current source, I_{CQ}. Since the capacitive reactance is very small, no AC voltage will appear across the capacitor, and it can be replaced by a battery of voltage V_{CC}, the voltage to which it is charged when no signal is present. With these two substitutions, the collector load circuit takes the equivalent form shown in Fig. 9-3*b*. Assume that a sinusoidal signal is present, and consider an instant of time when $i_C = 0$. At that instant, $i_L = I_{CQ}$, so that $v_{CE} = V_{CC} + i_L R_L$. From Fig. 9-5, $I_{CQ}R_L = i_L R_L = V_{CC}$, so that $v_{CE} = 2V_{CC}$. This establishes the upper limit on v_{CE}. When the signal reverses polarity, $i_C = 2I_{CQ}$. Then i_L must equal $-I_{CQ}$ so that $v_{CE} = 0$, establishing the lower limit.

With the amplifier biased as in Fig. 9-5, the currents and voltages of interest are

$$i_C = I_{CQ} + i_c = \frac{V_{CC}}{R_L} + i_c \tag{9.9}$$

$$i_L = -i_c \tag{9.10}$$

$$i_{\text{supply}} = i_L + i_C = I_{CQ} = \frac{V_{CC}}{R_L} \tag{9.11}$$

$$v_{CE} = V_{CC} - i_C R_L \tag{9.12}$$

$$v_L = +i_L R_L = -i_C R_L \tag{9.13}$$

If the signal current is sinusoidal, then

$$i_i = I_{i,\max} \sin \omega t \tag{9.14}$$

$$i_C = I_{C,\max} \sin \omega t \tag{9.15}$$

The maximum peak value of AC collector current is I_{CQ}; therefore

$$i_{c,\max} = I_{CQ} \sin \omega t \tag{9.16}$$

and

$$I_{c,\max} \leq I_{CQ} \tag{9.17}$$

The power supplied, the power dissipated in the collector and the load, and the efficiency are now found. The supplied power is

$$P_{CC} = V_{CC} I_{CQ} \approx \frac{V_{CC}^2}{R_L} \tag{9.18}$$

which is constant and essentially independent of signal current as long as the distortion is negligible. The power transferred to the load is

$$P_L = \frac{I_{i,\max}^2 R_L}{2} = \frac{I_{c,\max}^2 R_L}{2} \tag{9.19}$$

since $i_L = -i_c$, $I_{L,\max} = -I_{c,\max}$.

The maximum average power dissipated by the load occurs when

$$I_{c,\max} = I_{CQ} \tag{9.20}$$

so

$$P_{L,\max} = \frac{I_{CQ}^2 R_L}{2} = \frac{V_{CC}^2}{2R_L} \tag{9.21}$$

The collector dissipation is

$$P_C = P_{CC} - P_L = \frac{V_{CC}^2}{R_L} - \frac{I_{c,\max}^2 R_L}{2} \tag{9.22}$$

So the minimum power dissipated in the collector occurs when the maximum power is dissipated by the load:

$$P_{C,\min} = \frac{V_{CC}^2}{2R_L} \tag{9.23}$$

The maximum power dissipated by the collector occurs when no signal is present:

$$P_{C,\max} = \frac{V_{CC}^2}{R_L} = V_{CEQ} I_{CQ} \tag{9.24}$$

Then the efficiency for the inductor coupled amplifier with a sinusoidal signal is

$$Eff = \frac{P_L}{P_{CC}} = \frac{I^2_{c,\max}(R_L/2)}{V_{CC}I_{CQ}} = \frac{1}{2}\left(\frac{I_{c,\max}}{I_{CQ}}\right)^2 \tag{9.25}$$

From (9.25) it can be seen that the maximum efficiency will occur at maximum signal current and will be 50%. While an overall efficiency of 50% is not too desirable for power conversion purposes today, this analysis does provide some insight into the power operation of a class A power amplifier. Consider also that the efficiency would only be half as good given a resistor R_c in place of the choke in the DC collector circuit.

Class B Push–Pull Power Amplifier

As found in the previous discussion, the maximum attainable efficiency in class A operation is 50%, because the peak AC collector current never exceeds the quiescent collector current. In the class B amplifier, the DC collector current is less than the peak AC current. Thus less collector dissipation results, and the efficiency increases. Figure 9-6 is an example of a class B push–pull amplifier with a maximum efficiency of 78.5%, a great improvement over the class A amplifier discussed above.

Circuit operation is discussed, assuming ideal transistors, to determine the upper limits on efficiency and power output. Circuit operation can be explained in terms of the waveforms shown in Fig. 9-7. The center-tapped input transformer supplies two base currents of equal amplitudes but 180° out of phase. On the first half-cycle, i_{B_1} is zero, and because Q_1 is biased at cutoff, i_{C_1} is also zero. In this same interval i_{B_2} is positive, Q_2 conducts, and the collector current i_{C_2} is shown in Fig. 9-7*e*. So one transistor is cut off while the

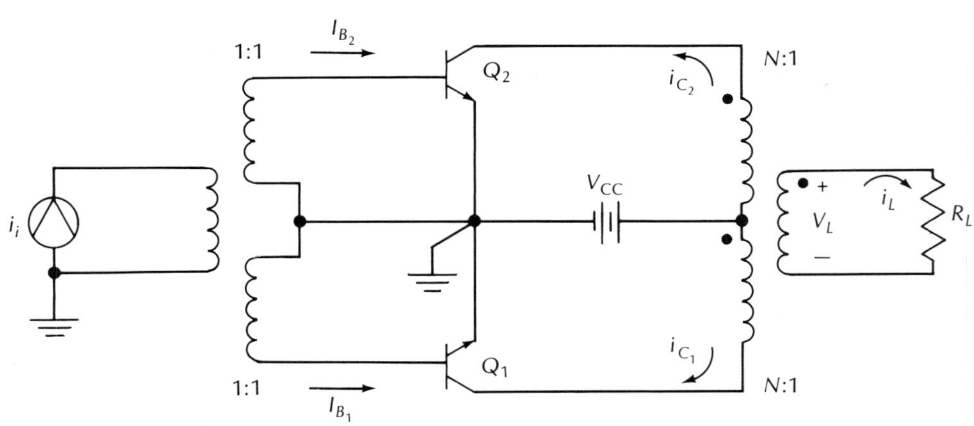

FIGURE 9-6 Class B push–pull amplifier.

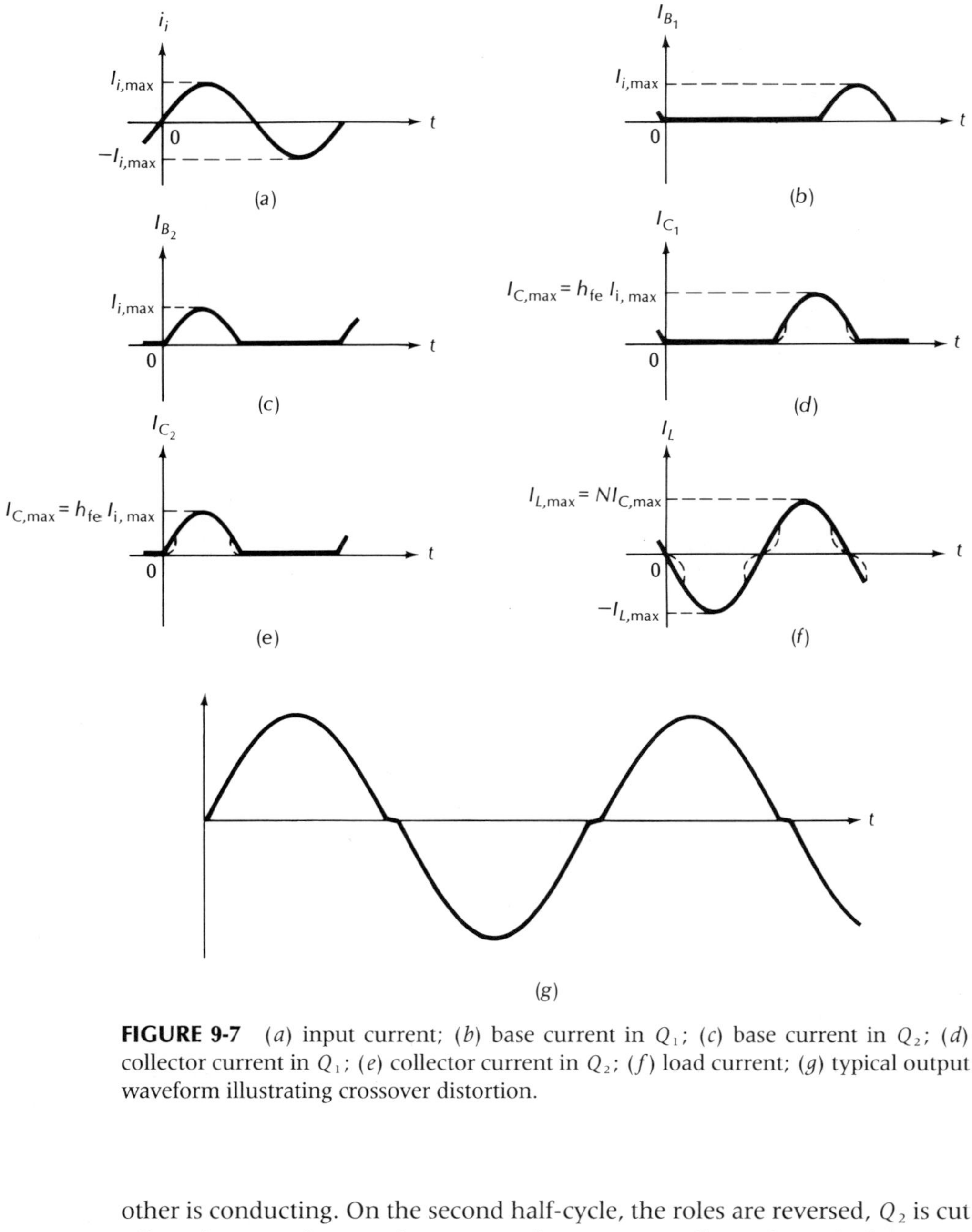

FIGURE 9-7 (*a*) input current; (*b*) base current in Q_1; (*c*) base current in Q_2; (*d*) collector current in Q_1; (*e*) collector current in Q_2; (*f*) load current; (*g*) typical output waveform illustrating crossover distortion.

other is conducting. On the second half-cycle, the roles are reversed, Q_2 is cut off, and Q_1 conducts. When Q_2 conducts, current flows through the upper half of the primary winding, and the resulting time varying flux in the transformer core induces a voltage in the secondary winding. This voltage, in turn, produces the first half-cycle of current through the load. When Q_1 conducts, the current i_{C_1} induces a flux in the core in a direction opposite to the flux of the previous half-cycle, resulting in the second half-cycle of load current. The final load current under these ideal conditions is thus directly proportional to

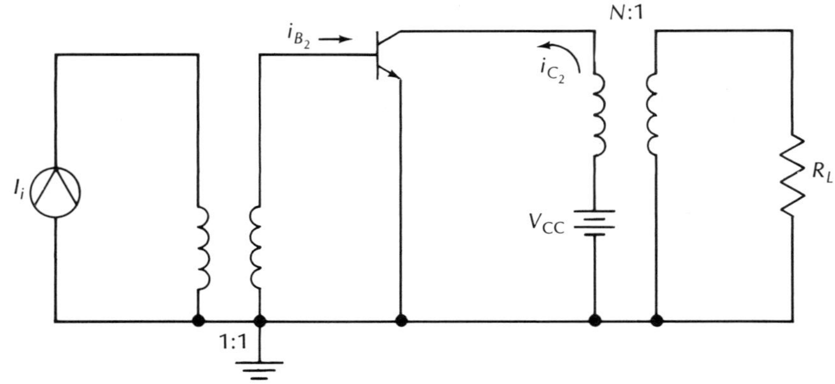

FIGURE 9-8 One-half of class B push–pull stage.

the signal current. It is seen from Fig. 9-8 that the load current i_L is related to the individual currents by

$$i_L = N(i_{C_1} - i_{C_2}) \tag{9.26}$$

If the circuit of Fig. 9-6 were used in practice, the load current would be extremely distorted near the zero crossing, as shown in Fig. 9-7*g*. This crossover distortion is due to the base-emitter voltage V_{be} being zero when no signal is applied. However, linear operation of the transistor begins only when i_B is positive enough for V_{BE} to exceed the cutin voltage, which is assumed to be approximately 0.65 V for silicon transistors. This distortion is represented as the dotted lines around zero crossing in Fig. 9-7. To eliminate this distortion, base–emitter junctions are biased at about 0.65 V. The result is then class AB, rather than class B operation, although it is so nearly class B that it is usually called simply "class B." Depending on the actual application, crossover distortion may be allowed and transformer or other stray capacitances will be counted on to filter it out. Since each transistor operates in a symmetrical fashion and only half the time, we need to study the operation of only one of the transistors. Consider Q_2 as shown in Fig. 9-8. This circuit enables us to describe the operation of the amplifier. The DC load line is a vertical line, $v_{CE} = V_{CC}$, and the AC load line has a slope

$$\frac{i_c}{v_{CE}} = -\frac{1}{R_L'} \tag{9.27}$$

as shown in Fig. 9-9. During the time that Q_2 is off, $i_{C_2} = 0$ and

$$v_{CE_2} = V_{CC} + Nv_L \tag{9.28}$$

varies from V_{CC} (which occurs when $v_{CE_1} = V_{CC}$, and hence $Nv_L = 0$) to $2V_{CC}$ (which occurs when $v_{CE_1} = 0$, and hence $Nv_L = V_{CC}$). Thus, while the tran-

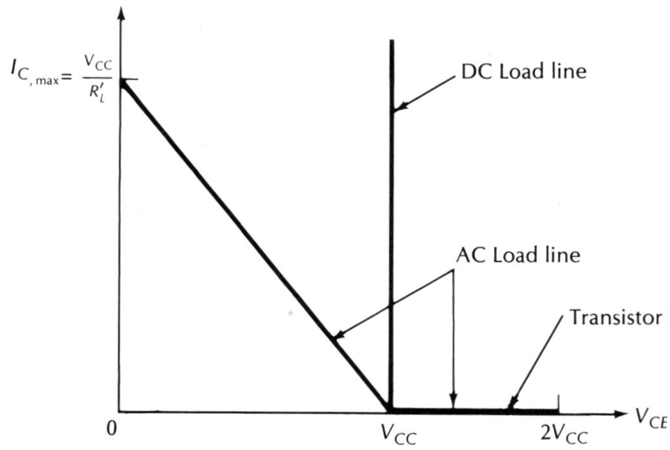

FIGURE 9-9 Load lines for class B stage.

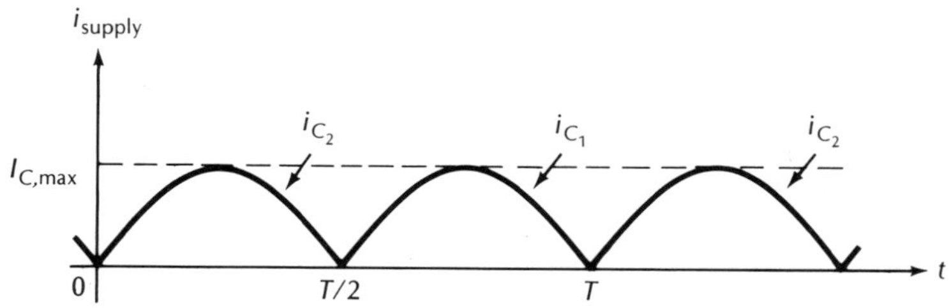

FIGURE 9-10 Power supply current waveform.

sistor is off, the AC load line is horizontal, with $i_{C_2} = 0$. The maximum value of both i_{C_1} and i_{C_2} is:

$$I_{c,\max} = \frac{V_{CC}}{R'_L} \tag{9.29}$$

Assuming that the signal current is sinusoidal, then (see also Fig. 9-10).

$$i_i = I_{i,\max} \sin \omega t \tag{9.30}$$

The power delivered by the supply is

$$P_{CC} = V_{CC} \frac{1}{T} \int_{-T/2}^{T/2} [i_{C_1}(t) + i_{C_2}(t)]\, dt \tag{9.31}$$

The current $i_{C_1} + i_{C_2}$ is the current flowing through the supply. This is full-wave rectified current, whose average value is $2/\pi$ times the peak value. Thus

$$\frac{1}{t} \int_{-T/2}^{T/2} (i_{C_1} + i_{C_2})\, dt = \frac{2}{\pi} I_{c,\max} \tag{9.32}$$

The supplied power P_{CC} is then

$$P_{CC} = \frac{2}{\pi} V_{CC} I_{c,\max} \tag{9.33}$$

Its maximum value is from Fig. 9-9:

$$P_{CC,\max} = \frac{2}{\pi} V_{CC} \frac{V_{CC}}{R'_L} = \frac{2V_{CC}^2}{\pi R'_L} \tag{9.34}$$

The power transferred to the load is

$$P_L = \frac{1}{2} I_{L,\max}^2 R_L = \frac{1}{2} I_{c,\max}^2 N^2 R_L = \frac{1}{2} I_{c,\max}^2 R'_L \tag{9.35}$$

Its maximum value is

$$P_{L,\max} = \frac{V_{CC}^2}{2R'_L} \tag{9.36}$$

The power dissipated in the collectors of transistors Q_1 and Q_2 totals

$$2P_C = P_{CC} - P_L \tag{9.37}$$

Using (9.33) and (9.35), we obtain

$$2P_C = \frac{2}{\pi} V_{CC} I_{c,\max} - \frac{R'_L I_{c,\max}^2}{2} \tag{9.38}$$

The maximum value of collector dissipation $P_{C,\max}$ is found by differentiating P_C with respect to I_{CM} and setting the result equal to zero:

$$2\frac{dP_c}{dI_{C,\max}} = \frac{2}{\pi} V_{CC} - R'_L I_{c,\max} = 0 \tag{9.39}$$

The collector current at which the collector dissipation is a maximum is then

$$I_{C,\max} = \frac{2}{\pi} \frac{V_{CC}}{R'_L} \tag{9.40}$$

and combining (9.38) and (9.40) gives the maximum collector dissipation:

$$2P_{C,\max} = \frac{2}{\pi} V_{CC} \frac{2V_{CC}}{\pi R'_L} - \frac{R'_L}{2}\left(\frac{2}{\pi}\frac{V_{CC}}{R'_L}\right)^2 = \frac{2}{\pi^2}\frac{V_{CC}^2}{R'_L} \tag{9.41}$$

The power dissipated in each collector is then

$$P_{C,\max} = \frac{1}{\pi^2}\frac{V_{CC}^2}{R'_L} \approx 0.1 \frac{V_{CC}^2}{R'_L} \tag{9.42}$$

The efficiency is calculated from (9.33) and (9.35):

$$Eff = \frac{P_2}{P_{CC}} = \frac{\frac{1}{2} R'_L I_{C,\max}^2}{(2/\pi) V_{CC} I_{C,\max}} = \frac{\pi}{4} \frac{I_{C,\max}}{V_{CC}/R'_L} \tag{9.43}$$

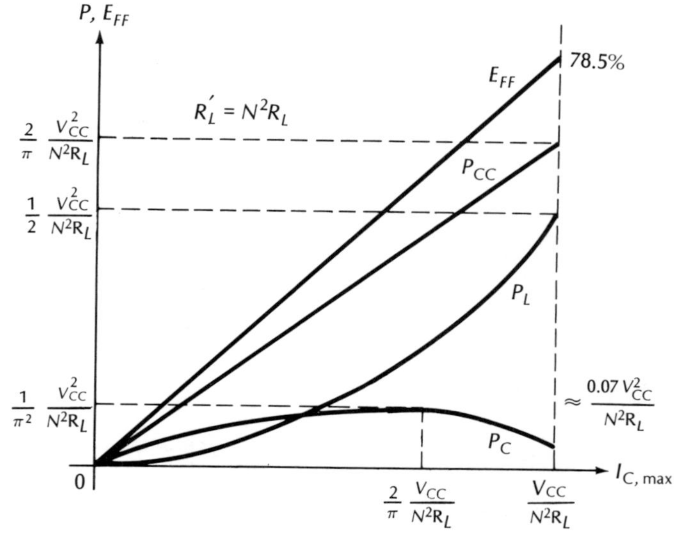

FIGURE 9-11 Power and efficiency variation in the class B push–pull amplifier.

Since the maximum attainable collector current is V_{CC}/R'_L, the maximum attainable efficiency is

$$Eff = \frac{\pi}{4} \approx 78.5\% \tag{9.44}$$

The supplied power, the load power, the collector dissipation, and the efficiency are plotted in Fig. 9-11. These results should be compared with those obtained using the class A amplifier discussed earlier.

Although it is true that the circuit requires two transistors and two transformers, the lower power rating of each of the transistors means that they will take up less space and require significantly less heat sinking than one high-power transistor, as in the class A amplifier. Another major advantage of the class B operation is that the zero signal current drain from the voltage source is zero, while in class A operation the zero signal current is the same as the full load current.

It is important to keep in mind that the efficiency and collector dissipation ratings are derived for sinusoidal signals and are theoretical maximums. Actual practice will yield efficiencies that only approach these numbers.

Complementary Symmetry

Another type of class B amplifier that requires no transformers is illustrated in Fig. 9-12. This type of amplifier employing one NPN and one PNP transistor uses complementary symmetry. When the signal voltage is positive, Q_1 con-

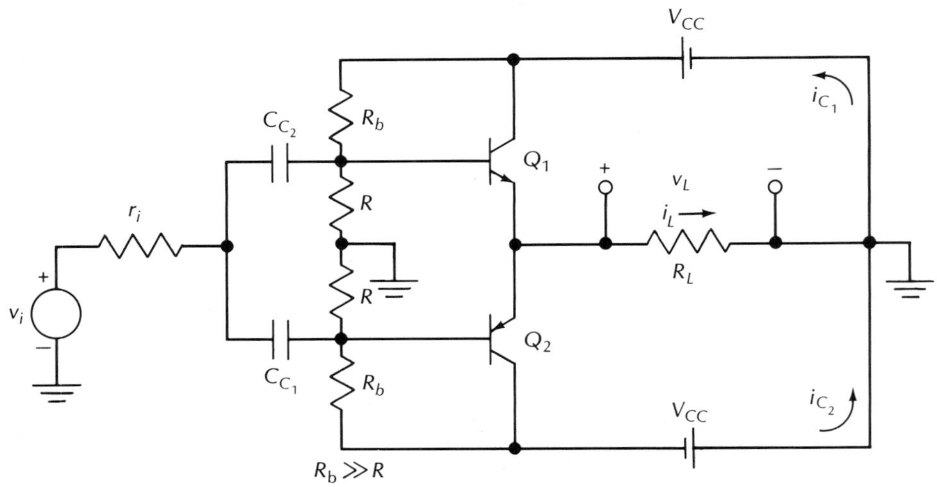

FIGURE 9-12 Complementary-symmetry amplifier.

ducts, while Q_2 is cut off. When the signal voltage is negative, Q_2 conducts while Q_1 is cut off. The load current is

$$i_L = i_{C_1} - i_{C_2} \tag{9.45}$$

The load line and output circuit power relations for this amplifier are the same as for the conventional class B amplifier discussed earlier. Some advantages of the circuit are that the transformerless operation saves on weight and cost and balanced push–pull input signals are not required. Disadvantages are the need for both positive and negative supply voltages and the problem of obtaining pairs of transistors matched closely enough to achieve low distortion.

The preceding discussions of class A and class B amplifiers was provided simply to illustrate the major difference between the two, where it matters most, in amplifiers of sufficient size to be of some use in powering critical equipment, such as computers. The difference in practical achievable efficiency will dictate the size, and cost, of large linear power amplifiers. Given all this, there are other alternatives to the linear approach that will provide even higher efficiencies. These will be discussed later on.

Having decided on a class B amplifier to provide the load power, you can take either of two approaches to the actual implementation. The first is very straightforward. Simply use the AC power line to supply the bulk DC voltage that the amplifier, in turn, converts to sine-wave voltage, at the amplitude and current required to run the load. Figure 9-13 shows an example of such a design. Figure 9-13 is a block diagram for a linear power amplifier. The preamplifier A_1 is used to amplify the input signal to a level that is adequate to drive the power amplifier A_2. The power amplifier A_2 provides the necessary sine wave signal to drive the output transformer T_2. The output transformer has isolated secondary windings that allow load isolation from the amplifier and oscillator circuitry. The power amplifier A_2 contains the overload and short-circuit protection circuitry. A local negative-feedback loop is taken from the output of the power amplifier back to the preamplifier, A_1. The overall negative feedback is taken from the feedback winding of T_2 back to the inverting input of the preamplifier and provides a closed-loop gain of 2.5 from the arm of R_1 to the primary of T_2. The positive current feedback signal is generated by sensing the *IR* drop across the primary of T_2 and applying this signal to transformer T_1. Then T_1 converts this differential signal into a single-ended signal and applies it to the input of the preamplifier through a resistive divider network containing the regulation control R_5. As this positive feedback is increased from zero with R_5, the output impedance of the power amplifier is reduced toward zero.

An alternate approach, which will lead to more AC input to output efficiency in the line regulator, is to use the power amplifier only to correct for the differences between the actual power-line waveform and a precision sine-wave reference built into the amplifier. This approach, illustrated in Fig. 9-14, requires less power delivered from the amplifier, for the same-size load, because the power line is still supplying most of the load current. The power amplifier itself is not necessarily any more efficient; it is simply supplying a smaller part of the overall total load power requirement.

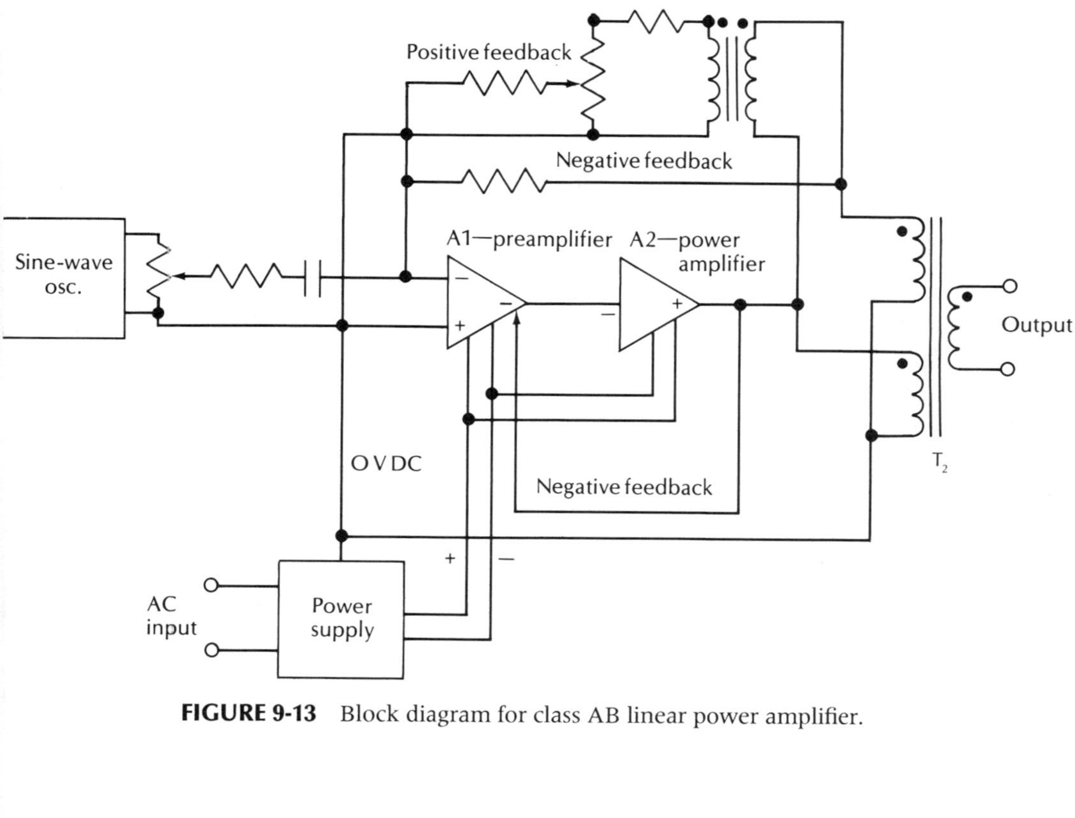

FIGURE 9-13 Block diagram for class AB linear power amplifier.

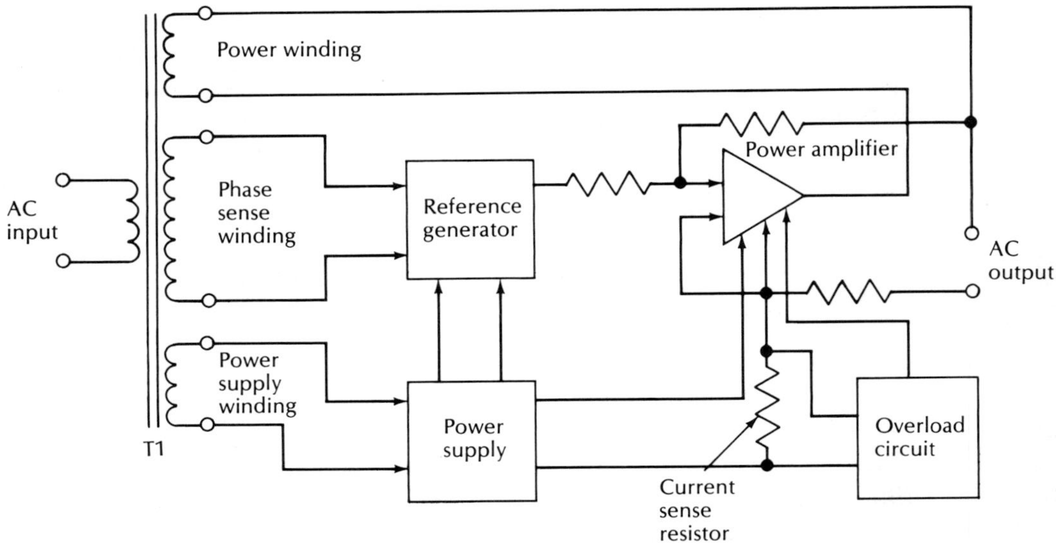

FIGURE 9-14 Block diagram of AC line regulator.

The circuit shown in Fig. 9-13 uses a transformer coupled output stage to step the sine-wave voltage from the power amplifier up to the voltage required to run the load. The turns ratio of the transformer will depend on the voltage supplied to the amplifier, since the amplifier is limited, when generating a sine wave, such that the RMS voltage it can deliver will be $0.707V_{peak}$. The voltage V_{peak} will be slightly less than the bulk voltage supplied to the amplifier because of the drop across the power transistors. If the bulk voltage is held low, the amplifier will be required to deliver more current, for a given load power, than if the bulk were high. If the bulk is high enough to make the peak voltage for a 115-V-AC sine wave, then no transformer is required on the amplifier output, and the currents in the amplifier will be held as low as possible. This becomes a decision based on the availability of high voltage power transistors, cost of same versus cost of the transformer, and considerations given for the generation of the bulk supplies.

Switching Regulator

A simple series pass linear regulator as shown in Fig. 9-15 is analogous to the linear power amplifier discussed earlier, when comparing the operation to a switching regulator. Figure 9-16 shows the same pass element now operating in saturation, rather than in its linear range. The improvement in operating efficiency and temperature are immediately obvious since the transistor is now either full on, saturated, or full off. With the linear regulator the power dissipated by the pass element is the voltage differential $(V_{input} - V_{output})I$. The same is true of the switching pass element, but when the transistor is operated in saturation, its input-to-output drop is likely to be only 1 V or less. With the

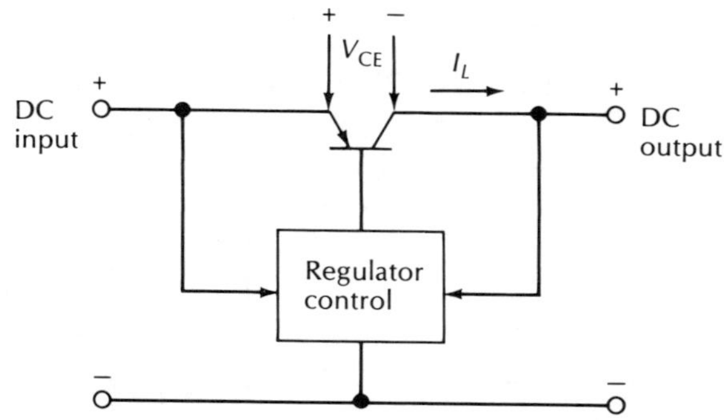

FIGURE 9-15 Linear DC regulator.

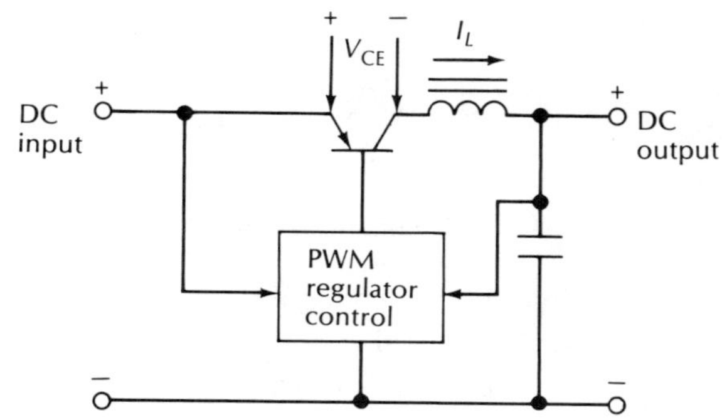

FIGURE 9-16 Switching DC regulator.

same current flowing this will result in much lower losses. The switching regulator also has switching losses, the losses associated with turning on and off, to deal with which adds to an overall reduction in efficiency. But compared with the linear regulator, a well-designed switching regulator will always be more efficient and have the potential for cooler operation and smaller packaging.

Using this analogy for comparison of linear versus switching AC power line regulators will reveal the same results. There are many ways to implement an AC power-line regulator utilizing switching power supply technology, and in all cases a well-designed circuit should outperform its linear counterpart every time. There will always be limitations to deal with. Linear power amplifiers will typically operate over a much broader frequency bandwidth than most designs utilizing switching technology. Since we are dealing only with power-line frequency here, however, that issue is of little importance.

Half-Bridge Converter

The circuit shown in Fig. 9-17 uses a half-bridge converter to synthesize a sine wave from two high-voltage DC rails. The AC power line is connected to a voltage doubler circuit made up of diodes CR_1 and CR_2, and capacitors C_1 and C_2.

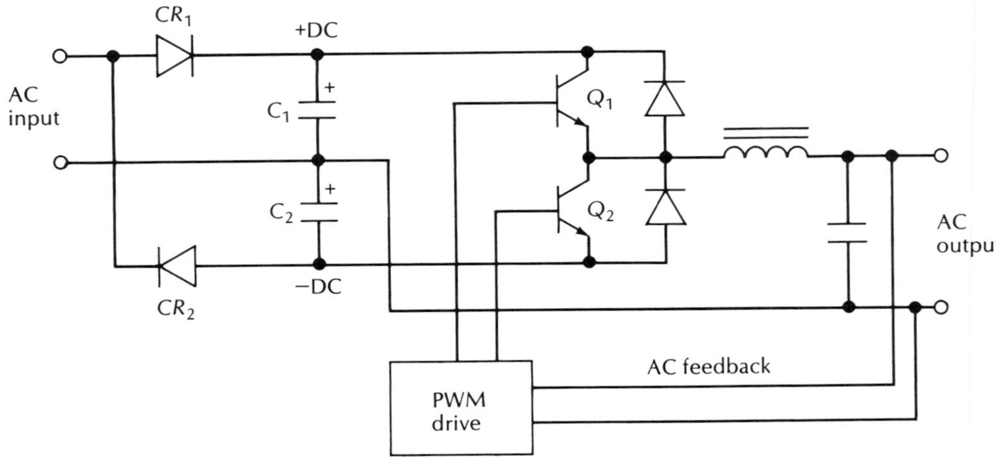

FIGURE 9-17 AC line regulator. Half-bridge configuration with PWM control.

Each capacitor will charge to the peak of the AC line voltage on alternate half-cycles, creating a plus and a minus rail, referenced to the center of the capacitors. If transistor Q_1 were to be turned on for one-half of the AC line cycle, and the Q_2 turned on for the other half, a square-wave voltage would appear across the load with a peak voltage approximately the same as the peak AC input line voltage. Now, instead of leaving Q_1 turned on for the entire half-cycle, it can be pulsed on and off for short intervals, over the half-cycle, followed by Q_2 pulsing in the same manner for the next half-cycle. If the pulses are arranged to provide the proper width, then the resultant output waveform will be a sine wave with a peak voltage approximately the same as that of the AC input line. This is discussed in more depth in a later chapter.

Once feedback regulation is accomplished with this circuit, it will cause the output voltage to remain constant while the input voltage varies. The difficulty here is that the peak of the output voltage can never be any higher than the peak of the input; therefore, while it will regulate the output voltage, preventing it from increasing as the line increases, it will not serve to regulate the output for decreasing input voltages. As the input AC line decreases, so will the DC level on the plus and minus rails, so the resultant AC output is limited in how much RMS voltage it can make, while still maintaining a sine wave. If it is important that the output waveform be maintained to a sine wave, then some other action must be taken. One possibility is to add an output transformer to the half-bridge. The output transformer can be an autotransformer as presented earlier, and its size will be quite small for the power rating of the regulator, given a boost of 10 or 20%. This will allow the power line to drop while the regulator maintains a constant output voltage. Another possibility is to place the transformer on the input rather than the output. By boosting the input voltage all the time, the plus and minus DC rails will be high enough to maintain the desired regulated voltage on the output, even during periods of low line. While either solution is acceptable, the addition of a line frequency transformer, even an autotransformer, adds additional size, cost, and weight. One disadvantage to adding the transformer ahead of the input is the added voltage burden placed on the power transistors in the DC-to-AC converter. If the transformer is on the output, then the transistors don't see the boost voltage. If maintaining a sine wave during low input line periods is not so important, then the simple solution is to modify the output waveform as required to maintain the desired RMS output. Another possibility, one that eliminates the need for the addition of low-frequency magnetics, is to add an additional boost converter between the input and the half-bridge converter. A simple high-frequency DC-to-DC converter operating off the rectified and filtered AC input can step the voltage up to sufficiently high plus and minus DC rails that the converter can maintain the desired output voltage over a very wide input range. Of course, added cost and size is still a consideration. These ideas are presented as possible solutions to the problem of maintaining the desired waveshape and voltage at the output of the regulator. It is up to the designer to determine what is actually necessary, given the load, and apply the best solution.

A feature of some line regulators, called *ride-through*, is a description of the regulators capability of continuing to provide an output to support the load for some period of time while the input power line goes to zero. Typically one or two cycles of ride-through, with full load applied, is specified by manufacturers of ferro-type line regulators, while tap-changing-type regulators will provide no ride-through at all. The ferro has some energy storage capability because of the nature of its design, so it does have an advantage in this area. The circuit shown in Fig. 9-17 also has ride through capability, and it is almost unlimited. The bulk capacitors that filter the input line and create the rails can be made as large as is desired to store energy when the input line disappears. The limitation here, of course, is the size and cost associated with a capacitor bank that will provide much more than a few cycles of ride-through. Eventually it becomes more economical to connect a battery across the capacitors and let that become the energy storage medium. Directly connecting batteries across the capacitor bus is possible, but not too practical, particularly in lower-power regulators where the cost of placing enough batteries in series to make up the rail voltages could drive the cost out of reach. Instead, it is more practical to use a lower voltage battery pack and a boost (DC-to-DC) converter, operating at high frequency to keep it small, to hold up the rails. Now the ride-through time can be determined by the size of the battery selected. Now we have just converted a line regulator into an uninterruptible power system. The difference is subtle since most UPSs are simple line regulators with lots or ride-through capability. The long ride-through is the difference in cost between a line regulator and a UPS.

This one circuit is presented here as an example of only one way to create a power-line regulator using switching techniques for high efficiency and small size. In Part IV, covering the topic of UPSs, more detail on the actual design of this and other topologies will be given. As you read into Part IV, keep in mind that many of the circuits presented can be applied to the power line regulator as well.

Load Considerations

Any time the load can be defined ahead of time, it makes the ultimate design of the line regulator easier. Whatever the load, it will rarely display a unity power factor, with sine-wave current. Instead it will either display sine-wave current that either leads or lags the voltage, as is the case with displacement power factor, or the current will be in phase with the voltage, but be very nonlinear, displaying a large crest factor.

Displacement Load Power Factor

The displacement load power factor must be considered when designing a linear power amplifier. The power required to run a load will depend on the elements making up the load. All equipment that operates off the power line will present some degree of series or parallel resistance R and reactive element, inductance L and capacitance C. Therefore, the total power to be delivered to the load by the amplifier will be

$$P = EIpf \tag{9.46}$$

where pf is the cosine of the phase-angle difference between the voltage across the load and the current through the load, expressed as the power factor. Figure 9-18 illustrates the relationship of a sine-wave voltage and sine-wave current where a power factor load is involved.

In an ideal purely capacitive circuit the current would lead the voltage by 90°, while in a purely inductive circuit the current would lag the voltage by 90°. Since it is impossible to achieve purely inductive or capacitive elements in the real world, all loads will have some of each, in addition to some resistance. In a purely resistive load the total load power would simply be the

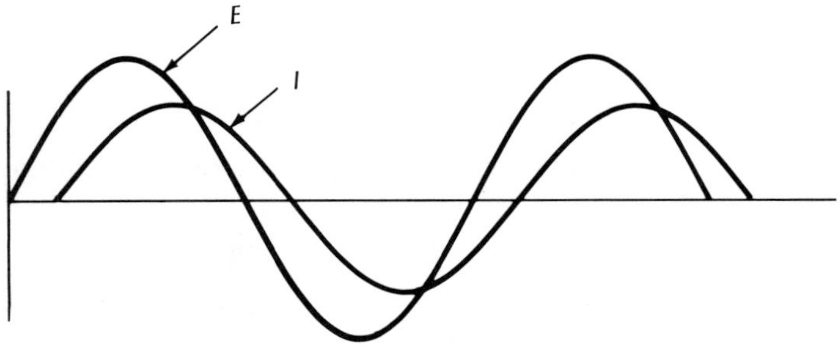

FIGURE 9-18 Voltage and current waveforms in displacement power factor load.

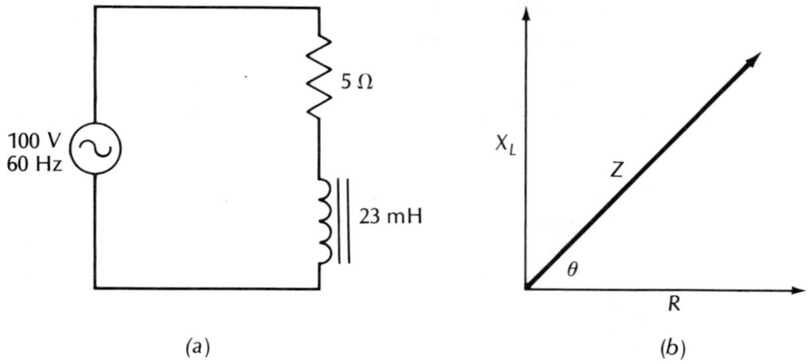

FIGURE 9-19 (*a*) Series *RL* circuit; (*b*) vector representation.

load voltage multiplied by the load current, referred to as *VA*. In this case *VA* will equal the wattage (W). From (9.46) it is shown, though, that any amount of phase displacement or power factor will result in the true power in the load being less than the measured *VA*. This is because the maximum phase angle associated with any load will be no more than 90°, leading or lagging. The cosine of any angle between −90° and +90° will be a number between 0 and 1. The cosine of 90° is 0, while the cosine of 0 degrees is 1. Therefore, the power in a purely capacitive load would be $VA \times 0 = 0$, while the power in a purely resistive load is $VA \times 1 = VA$. Figure 9-19 shows an example of a circuit containing resistance and inductance as the primary load elements, along with a vector representation of the load. The power and power factor for this circuit can be solved as follows:

$$X_L = 2\pi fL = 377 \times 23 \times 10^{-3} = 8.67\ \Omega \tag{9.47}$$

$$Z = \sqrt{R^2 + X_L^2} = 10\ \Omega \tag{9.48}$$

$$\theta = \text{arc tan}\ \frac{X_L}{R} = 60^\circ \tag{9.49}$$

$$Pf = \cos\theta = \cos 60^\circ = 0.5 \tag{9.50}$$

Then the total load

$$I = \frac{E}{Z} = \frac{100}{10} = 10\ \text{A} \tag{9.51}$$

$$P_{\text{total}} = EI = 100 \times 10 = 1000\,VA \tag{9.52}$$

$$P_R = I^2R = 10^2 \times 5 = 500\ \text{W} \tag{9.53}$$

$$P_L = I^2X_L = 10^2 \times 8.67 = 867\ \text{VARS} \tag{9.54}$$

A difficulty you encounter regarding where power factor loads are connected to linear power amplifiers is increased dissipation in the power transistor. An

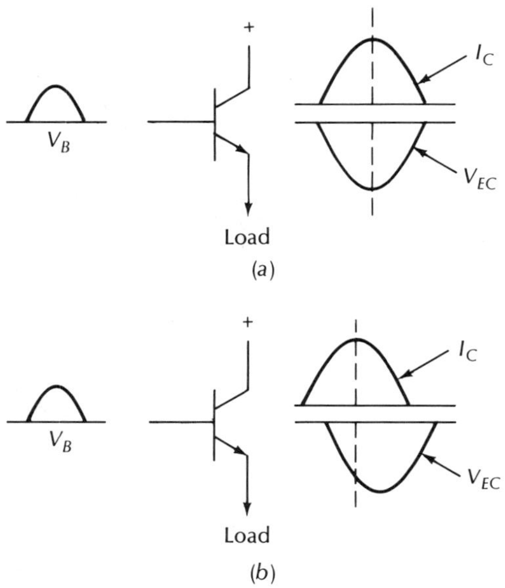

FIGURE 9-20 (*a*) Unity Pf load current reaches maximum when voltage across transistor is minimum; (*b*) phase-displaced load current reaches maximum when voltage across transistor is still high.

example of this is illustrated in Fig. 9-20. Since the transistor is operating in its linear region, the voltage across the part V_{CE} and the current through the part I_C will be sinusoidal. As the transistor is driven harder through the full half-cycle, the voltage across the part reduces, effectively pulling the emitter voltage up to near the collector voltage at the peak of the half-cycle. In Fig. 9-20*a*, where the load power factor is unity, *I*yC is in phase with V_{CE}, and therefore the peak current through the transistor occurs at the same time as V_{CE} is at a minimum. By changing from a unity power factor load to a load with some phase displacement, as shown in Fig. 9-20*b*, the peak current through the transistor now occurs at a time when V_{CE} is not at its minimum. The result is higher power dissipation in the transistor, for the same given amount of load current.

Crest Factor Load

Figure 9-21 illustrates what is known as a *high-crest-factor current waveform*. As shown, the peak of the current waveform is approximately three times the RMS value. A purely sinusoidal current waveform would have a peak to RMS ratio of only 1.414 by comparison. This is the type of waveform typically found when operating a switching power supply that has not had any type of power factor correction circuitry added to prevent this high current peak. The cause of the high current peak is the capacitor found at the front end of the supply. The AC power-line voltage is rectified and filtered by the capacitor. The capacitor is sized for the load the power supply is supposed to run, so

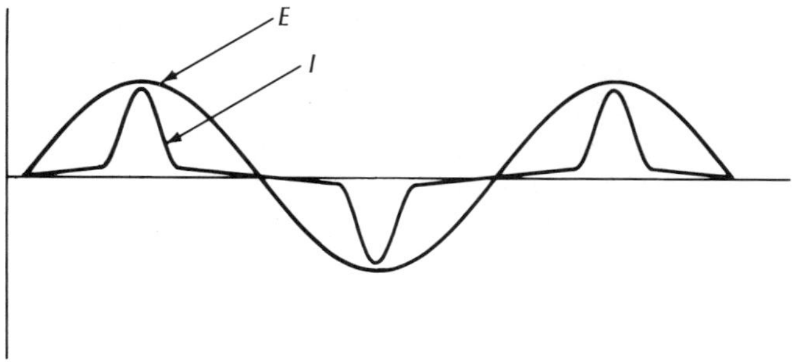

FIGURE 9-21 High-crest-factor current waveform compared to voltage waveform.

that between voltage peaks the DC level of the voltage on the capacitor will drop only a slight amount. This means that the capacitor is still near full charge at the start of the next half-cycle. So when the AC voltage fed to the capacitor increases, only slight current will flow until suddenly the AC voltage goes above the DC voltage on the capacitor. This causes a very fast rising current to flow into the capacitor while it is charging up to the peak of the AC voltage.

The power factor caused by this sort of activity will be on the order of 0.65. Using $VA \times pf$ to calculate the true load power drawn by the switching power supply will reveal one of the problems associated with this type of load. A typical switching power supply that draws a measured 500 W of power from the power line when fully loaded will have to be drawing an RMS current on the order of 6.5 A. But a resistive load of 500 W will draw only 4.2 A from a 120-V-AC line. This extra 2.3 A must be supplied to operate the power supply correctly, and must be taken into account when designing a regulator that is expected to operate into nonlinear loads, such as switching power supplies.

Summary

When considering regulating the AC power line to protect against the wide variations that can be experienced, many possibilities exist. In this chapter we have dealt with regulators that are very simple and easy to implement to more complex linear and switching power amplifiers. There is no simple rule or method for determining which solution is best for the application at hand. When the load can be well defined and you are designing a regulator for that load, there may be several solutions to choose from. When designing a regulator that will be expected to operate any given load within its power range, then the selection is narrowed down. The amount of regulation is also an important consideration when deciding which topology to use. The tap changer, for example, is much simpler in concept and more cost-effective than a linear or switching power amplifier with the same power-handling capability. If the application requires precise regulation and extremely fast response time to load or line fluctuations, however, a linear or switching amplifier is a better choice. Many companies that provide linear power amplifiers for use in automated test systems or other applications are beginning to use switching techniques to improve on cost, package size, and so on. But there are still some limitations that the linear amplifier overcomes, and it still has a place in the AC power conversion business.

PART IV

Blackout Protection

Studies have shown that of all the various problem associated with the power line, the blackout, or complete loss of utility power, is likely to happen the least. Even when it does happen, if the cause was external to the building, it is most likely to be proceeded by any or all of the other anomalies described in the earlier chapters. Surges, radio interference, and even brownouts are all more common and easier to deal with than blackouts. With all the other forms of power problems we are simply somehow filtering or otherwise manipulating the existing voltage to make it better suited to the load. But a blackout leaves us with nothing to work with; therefore, action must be taken to generate power that can replace the primary source in the event that it fails. To do this requires some form of energy storage, in the form of either batteries or rotating mass such as a flywheel and then some method of converting that energy into useful power to operate the load. And, of course, while the power line is present it must be used to keep the source of the stored energy topped off. All this is accomplished in modern power protection equipment called *uninterruptible power systems* (UPS) or *standby power systems* (SPS). These systems take on many forms, utilizing an array of technologies, to ensure that when the power line fails the load doesn't know anything about it. Both SPS and UPS systems can utilize all the same technologies: the primary difference between them is the application. Later we will deal more with the actual application and discuss the differences, both advantageous and disadvantageous, between SPS and UPS designs.

As mentioned above, one possible method of storing an emergency source of energy is with a flywheel. Several manufacturers of power conditioning equipment on the market utilize motor-generator sets to provide load power. The motor receives its energy from the power line and, in turn, runs the generator, which puts out clean AC. If the power line should fail, the motor will continue to rotate on its own flywheel energy long enough to allow a battery to switch in and continue as the source of energy to keep the motor operating. These systems are reliable and very effective but are expensive to implement, particularly at low power levels. For this reason, they are mentioned here since they are one of the choices in power protection products, but they are not discussed further. We are more interested in solid state power conversion and will attempt to provide a well rounded discussion of the principles and topologies in use today. This discussion will begin with an overview of some basic DC-to-DC switching converters. The principles of DC-to-AC conversion are very similar, but additional considerations must be brought out. Discussions of filter design, thermal management, and batteries will complete this section.

CHAPTER 10
Switching Converter

A switching converter is generally utilized to convert a voltage from one DC level to another. Even what is referred to as an *offline* switching power supply first rectifies the AC voltage and filters it before the switching converter processes it to the final desired voltage(s). This chapter, as well as those following, is concerned with the conversion of DC voltage to AC voltage, utilizing solid-state switching techniques. The term "switching" implies that the power devices in the converter, whether bipolar junction transistors or power MOSFET (metal oxide semiconductor field-effect transistor), are being operated at or near saturation as an attempt to minimize forward voltage drop in the device, thus minimizing losses. The following discussion of switching converters will cover the switching characteristics of bipolar and MOSFET transistor, as well as various topologies to effect the desired conversion.

Bipolar Transistor Switching Characteristics

The transistor, when operated in its linear or active region, will respond to the input base signal with little concern for delay. When operated in saturation, as is the case when it is used as a switch, then the response time is of concern. Delays in turn-on or turn-off of the switch will have an effect on the maximum operating frequency at which the converter can operate. For example, we can't have the pulses to the base of a transistor so close together that the transistor never has time to turn off between pulses, or so narrow that the transistor never turns on.

The transistor can be made to operate as a switch by designing the associated circuit so that the transistor is in either the cutoff or saturation region. When the transistor is in saturation, maximum collector current flows and the transistor is turned on. The transistor is controlled by the current applied to the base. Such a transistor switch is shown in Fig. 10-1. The response of the transistor switch to a pulse-waveform input voltage v_i is shown in Fig. 10-2.

Calculations of the delay time, rise time, fall time, and storage time (described in the following list) involve many approximations and generally result in values that differ significantly from those published by the transistor's

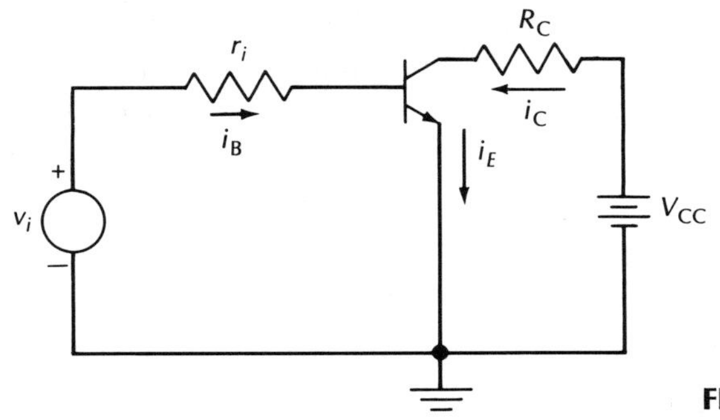

FIGURE 10-1 Transistor switch.

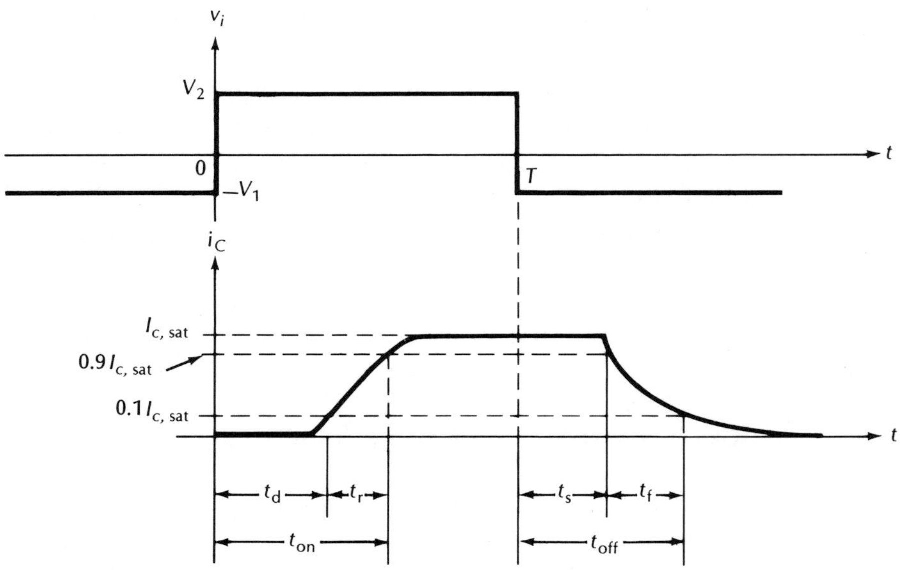

FIGURE 10-2 Response of a transistor switch to a voltage pulse.

manufacturer. Seldom do you find the published specifications for switching times at the same operating voltage and collector current that you intend to operate the transistor in your circuit. For this reason it is important to use the published specification as a guide, but make measurements of the actual part in a test fixture before making final design decisions.

Turn-on-time. There is a delay between the leading edge of the input voltage pulse and the time that the collector current takes to reach 90% of its maximum value. This time is called the *turn-on time* t_{on}. It is divided into two time intervals; the first is called the *delay time* t_d, and the second, the *rise time* t_r.

Delay time. The delay time is the time required for the collector current to increase to $0.1I_{C,\text{sat}}$. Another way of describing t_d is to note that it is ≈ equal to the time required for the base-to-emitter diode voltage to increase from $-V_1$ to ≈ +0.7 V.

Rise time. The rise time is the time required for the collector current to increase from $0.1I_{C,\text{sat}}$ to $0.9I_{C,\text{sat}}$. During this interval the transistor is operating in the normal active region.

Turn-off time. The time required for the collector current to decrease from $I_{C,\text{sat}}$ to $0.1I_{C,\text{sat}}$ when v_i goes negative is called the turn-off time t_{off}. The turn-off time is the sum of the storage time t_s and the fall time t_f, as shown in Fig. 10-2.

Storage time. The storage time is the elapsed time from the trailing edge of the input pulse ($t = T$) to the point where i_C just starts to decrease toward zero.

Fall time. The fall time t_f is the time it takes for the collector current to decrease from $I_{C,\text{sat}}$ to $0.1I_{C,\text{sat}}$.

Propagation delay time. The time it takes for the transistor switch to respond to an input signal is called the *propagation delay time* t_{pd}. This is usually defined as the time between the midpoints of the input-signal transition voltage and the output-signal transition voltage, as shown in Fig. 10-3.

The propagation delay when the switch output falls from the high state to the low state is symbolized as $t_{\text{pd,hl}}$ or $t_{\text{pd}-}$. Similarly, the propagation delay when the switch output rises from the low state to the high state is called $t_{\text{pd,lh}}$

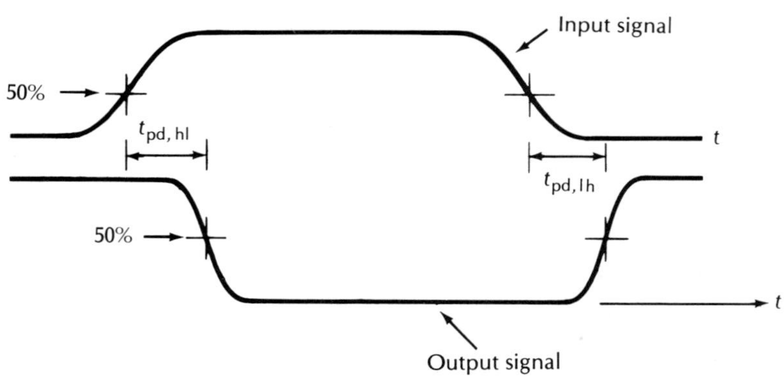

FIGURE 10-3 Propagation delay time.

or t_{pd+}. Usually t_{pd+} is greater than t_{pd-} because of the inevitable capacitance at the switch output. When the output falls, the capacitance is discharged through the transistor, which acts like a low impedance; while when the output rises, the capacitor is charged through the load resistor, a much higher impedance, and therefore the charging time is longer. If a complementary-symmetry output stage is used, the propagation delay times are comparable.

When a transistor is saturated, excess charge is stored in the base–emitter and base–collector junctions. If we attempt to suddenly reverse-bias a diode, driving the diode into cutoff, we would first have to discharge the diode storage capacitance C_s. Since the base–emitter and base–collector junctions of a transistor can be considered to be forward-biased diodes, we expect that in order to drive a saturated transistor to cutoff we must first remove the excess charge stored in the base–emitter and base–collector diode capacitances. A Schottky diode connected between base and collector serves to speed the transition from forward bias to cutoff since in a Schottky diode C_s is approximately zero. The Schottky diode can also be used to prevent a transistor from saturating. Such a configuration, consisting of a Schottky diode and a transistor is shown in Fig. 10-4.

When the Schottky diode is placed across the transistor, the base-collector voltage of the transistor cannot become more positive than 0.3 V, the forward voltage of the Schottky diode. Since $v_{CE} = v_{BE} - v_{BC}$, approximately $0.7 - 0.3 = 0.4$ V, we see that the collector–emitter voltage of the transistor will always be greater than or equal to 0.4 V. This is but one of many possible antisaturation techniques available. There are obvious advantages to operating the transistor in saturation, thus minimizing the collector–emitter voltage

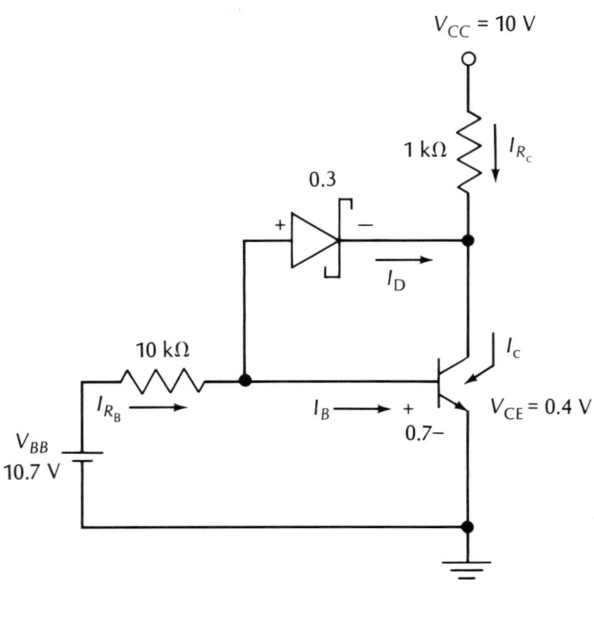

FIGURE 10-4 Effect of Schottky diode on the quiescent operation of a transistor amplifier.

drop. Operating the transistor in quasi-saturation by applying an antisaturation drive will reduce the turn-off delay caused by saturation, and the trade-off will yield more efficient operation.

Bipolar Power Transistor Considerations

In power transistors, the main design consideration is power-handling capability. This capability is determined by the maximum junction temperature a transistor can withstand and how quickly the heat can be conducted away from the junction. In general, the basic physical theory that defines the behavior of any bipolar transistor in relation to charge–carrier interactions, current gain, frequency capabilities, voltage breakdown, and current and temperature ratings is not significantly different for power types. Power transistors, however, must be capable of large current densities and are required to sustain large voltages. For power transistors, therefore, the basic theory must be expanded to include the effect that these conditions have on the physical behavior of the devices. In addition, the physical capabilities of power transistors must be defined in terms of factors such as second-breakdown energy levels, safe operating area, and thermal-cycling stresses, which are not usually considered for small-signal types.

The currents in a transistor are directly related to the movement of minority carriers in the base region that results from the application of voltages of the proper polarities to the emitter–base junctions. A definite mutual relationship exists between the transistor currents and the voltages applied to the transistor terminals. Graphical representations of the variations in transistor currents with the applied voltages provide and excellent indication of the operation of a transistor under different biasing conditions. Transistor manufacturers usually provide curves of current–voltage characteristics to define the operating characteristics of their devices. Such curves are provided for either common-emitter or common-base connections. Figure 10-5 shows the bias-voltage po-

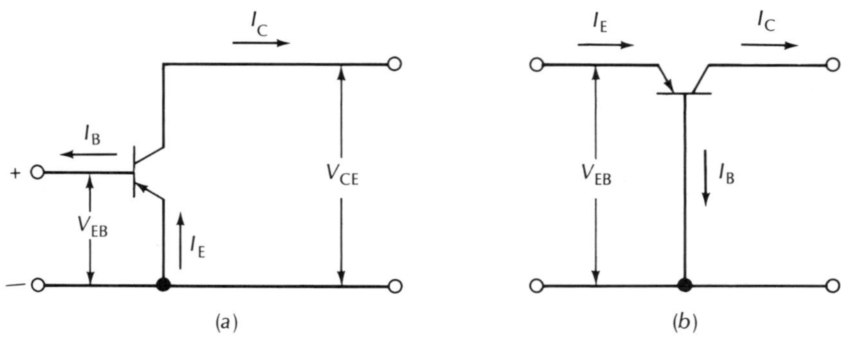

FIGURE 10-5 Transistor bias voltage polarities: (*a*) common-emitter connection; (*b*) common-base connection.

larities and the current components for both common-emitter and common-base connections of a *PNP* transistor. For an *NPN* transistor, the polarities of the voltages and the directions of the currents are reversed.

The common-emitter connection, shown in Fig. 10-5*a*, is more widely used in practical applications. In this connection, the emitter is the common point between the input (base) and the output (collector) circuits, and large current gains are realized by use of a small base current to control a much larger emitter-to-collector current. The common-base connection (Fig. 10-5*b*), differs from the common-emitter connection in that the voltages applied to the transistor are referred to the base rather than to the emitter.

Common-Base Characteristics

Figure 10-6 shows the input and output current–voltage characteristics of a typical *PNP* transistor in a common-base connection. The input characteristic curves in Fig. 10-6*a* show the base current as a function of the emitter–base voltage for different values of collector–base voltage. For any given value of collector voltage, the base current varies with the emitter–base voltage in a manner similar to that of any forward biased narrow *pn* junction. After the initial interval required for the forward bias voltage to overcome the energy barrier at the junction, the current to voltage relationship is almost linear. The slight effect that the collector voltage has on the base current results because variations in the collector voltage change the effective width of the base.

The effective base width is the distance between the edges of the opposing (collector and emitter) depletion layers. This distance is significantly less than the actual thickness of the *n*-type base material because both depletion layers penetrate into the base region. Because of the small forward bias applied to the emitter–base junction, the emitter depletion layer is much narrower than that of the collector–base junction, and it does not change appreciably with

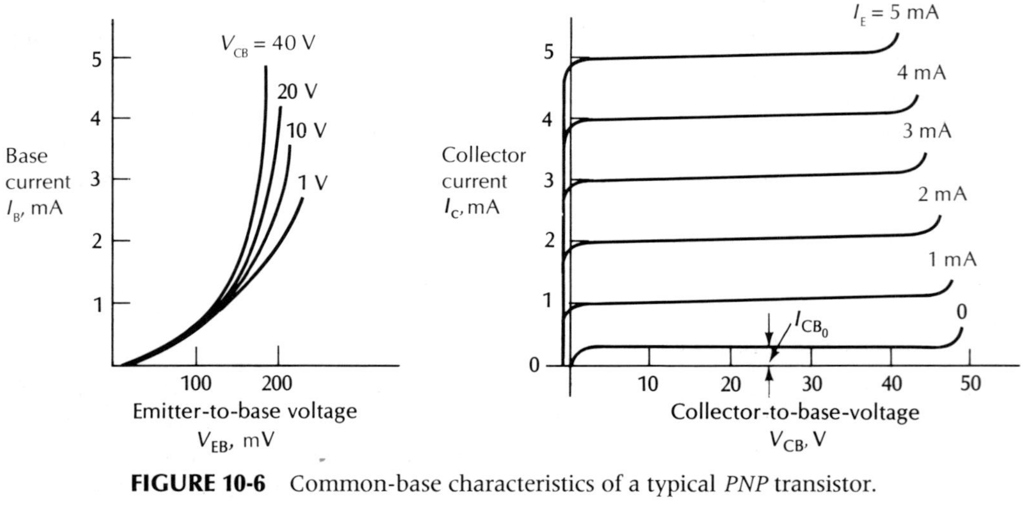

FIGURE 10-6 Common-base characteristics of a typical *PNP* transistor.

the small range of variation allowed in the emitter-to-base voltage. The collector junction depletion layer, however, varies markedly with the much larger changes in collector voltage and causes an attendant change in the effective base width. An increase in collector voltage causes a slight increase in base current (for a constant emitter-to-base voltage) because the hole gradient in the *PNP* transistor increases as the effective base width decreases. Many other transistor characteristics are also affected by changes in collector voltage because of the dependence of the effective base width on the collector depletion layer.

The common-base output characteristics shown in Fig. 10-6*b* reveal that the collector current is very nearly equal to the emitter current and is largely independent of collector voltage. This current is made up of the small reverse current I_{CBO} which results from the extraction of thermally generated holes from the base region, and the forward current produced by the diffusion of holes from the emitter. The diffusion current is almost independent of voltage because all the holes that reach the edge of the collector depletion layer are extracted by the reverse-biased junction. The collector current remains essentially constant, even down to zero voltage, at which point the excess holes are still extracted by the collector. A small forward voltage (< 1 V) must be applied to the collector to increase the hole density just outside the junction and oppose the diffusion from the emitter in order to reduce the collector current to zero. The characteristic curves show that the collector current varies extremely rapidly with voltage in this region. The collector reverse leakage current I_{CBO} is measured with the emitter circuit open. The presence of the second junction, however, still affects the level of this current because the emitter acquires a small negative bias when it is open-circuited. This bias reduces the hole gradient at the collector and causes the reverse current to decrease. This current is much smaller with the emitter open than when the emitter–base junction is short circuited. In the characteristic curves shown in Fig. 10-6*b*, the magnitude of the current I_{CBO} is considerably exaggerated. This current is normally in the order of microamperes or less, although it is often increased by excess surface currents in the same way as the reverse current of any *pn* junction. The reverse current increases with collector voltage and may lead to avalanche breakdown at high voltages. All of the collector characteristic curves bend upward at sufficiently high collector voltages because of multiplication in the collector depletion layer. The multiplication factor depends on the junction voltage only; the curvature of the characteristics, therefore, is more noticeable for high collector current.

Common-Emitter Characteristics

The common-emitter characteristics of a *PNP* transistor shown in Fig. 10-7 are similar to those of the common-base connection. There are several important differences; however, the most marked is the small magnitude of the base current that replaces the emitter current in both sets of curves. The base current consists of two components, each of the order of microamperes. These components include the current produced by an inward flow of electrons to replace those lost in the hole injection and diffusion mechanisms, and a current that is largely the result of the extraction of thermally generated holes making up the leakage current of the collector. The total base current, therefore, is just sufficient to make up the losses in the current transfer from emitter to collector. If the base current were increased by means of the external circuit, more holes could diffuse from emitter to collector, and a considerable increase in the collector current would occur. Physically, this condition is produced by increased hole injection and, therefore, by an increase in the emitter-to-base voltage required to produce and increase in base current. Although changes in base current and voltage must occur together, it is preferable to think of the transistor as a current-controlled device; for this reason, the common-emitter characteristics are usually shown for constant base currents, as indicated in Fig. 10-7. The curves in this figure confirm that small changes in base current produce much larger changes in collector current.

Another feature of the common-emitter output characteristics is that the common-emitter reverse collector current I_{CEO}, measured with zero input current ($I_B = 0$ in this case), is much larger than the reverse collector current I_{CBO} in the common-base connection. When the base current is zero, the emitter current adjusts itself so that the losses in the hole-injection and diffusion mechanisms are exactly balanced by the supply of excess electrons left in the vicinity of the collector by hole extraction. For this condition, the collector current is equal to the emitter current. The common-emitter reverse collector current I_{CEO} increases with collector voltage, unlike the common-base reverse collector current I_{CBO}. This behavior is another consequence of the variation in the effective base width with collector voltage. The narrower

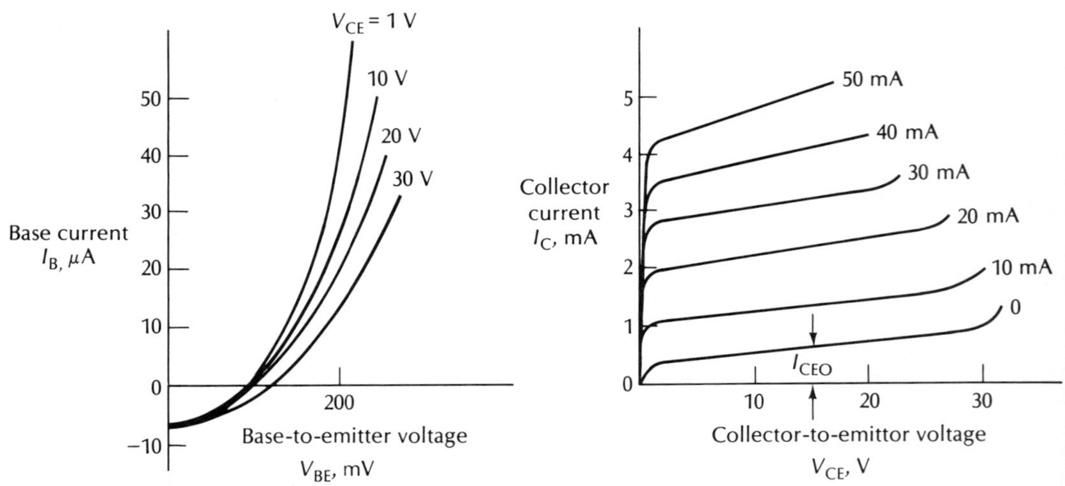

FIGURE 10-7 Common-emitter characteristics of a typical *PNP* transistor.

the effective base region, the more efficient is the transfer of current from emitter to collector. The more efficient base transport with the higher collector voltage permits a higher emitter current to flow before the losses are again balanced by the supply of electrons from the vicinity of the collector. A final point to note about the common-emitter characteristics is that the current falls rapidly to zero for small collector to emitter voltages. The collector-to-emitter voltage is normally divided between the two junctions to provide a small forward bias for the emitter–base junction and a much larger reverse bias for the collector–base junction. If the emitter to collector voltage is reduced to the "saturation" value, which is a fraction of a volt, the collector reaches zero bias; below this value, it becomes slightly forward-biased. This condition occurs near the knee of the characteristics. Slightly below the saturated condition, holes continue to flow into the collector, in spite of its forward bias. If the emitter-to-collector voltage is reduced further, the collector current falls rapidly to zero. The input characteristics for the common-emitter connection are similar to those for the common-base connection, except for the greatly reduced magnitude of the base current compared with the emitter current. For the base current to be zero, the base–emitter junction must have a small forward bias; as a result, all the input characteristics cross the current axis. The collector voltage again affects the characteristics by varying the effective base width, and thus the current gain. With a constant emitter hole density (V_{BE} constant), the base current falls as the collector voltage is raised, and the base transport becomes more efficient.

Effects of Temperature on Bipolar Transistor Characteristics

Transistor characteristics will vary with temperature changes. Since most circuits must operate over a wide range of environments, a good circuit design should compensate for such changes so that operation is not adversely affected by the temperature change.

Current Gain

The effect of temperature on the gain of a silicon transistor is dependent on the level of the collector current, as shown in Fig. 10-8. At the lower current levels, the current-gain parameter h_{FE} increases with temperature. At higher currents, however, h_{FE} may increase or decrease with a rise in temperature because it is a complex function of many components.

Base-to-Emitter Voltage

Figure 10-9 shows the effect of changes in temperature on the base-to-emitter voltage V_{BE} of silicon transistors. Two factors, the base resistance r'_{bb} and the height of the potential barrier at the base–emitter junction V'_{BE} influence the behavior of the base-to-emitter voltage. As the temperature rises, material resistivity increases; as a result, the value of the base resistance r'_{bb} becomes greater. The barrier potential V'_{BE} of the base–emitter junction, however, de-

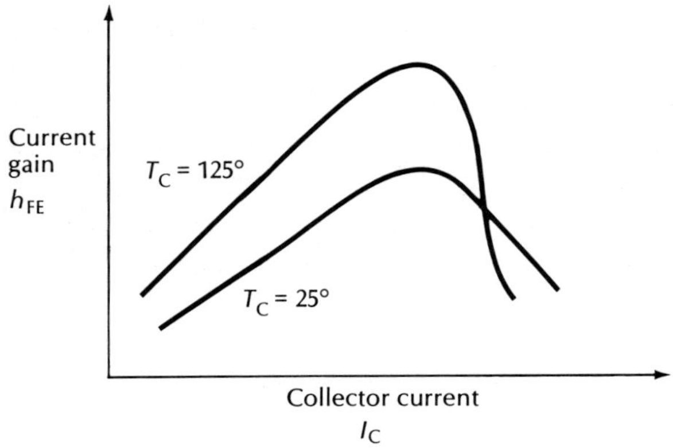

FIGURE 10-8 Current gain as a function of collector current at different temperatures.

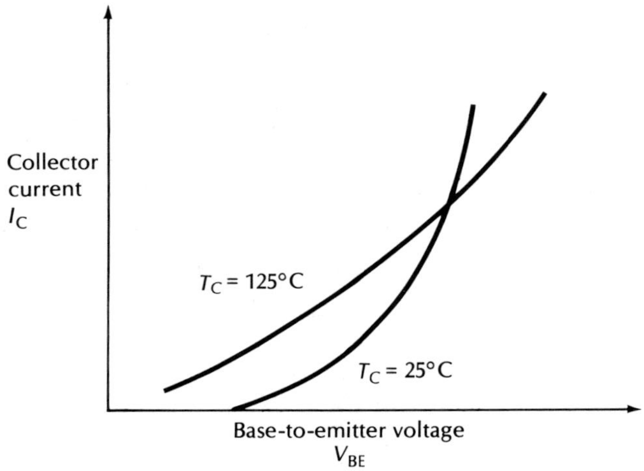

FIGURE 10-9 Collector current as a function of base-to-emitter voltage at different temperatures.

creases with temperature. Equations 10.1 and 10.2 show the relationship between the base-to-emitter voltage and the two temperature-dependent factors:

$$V_{BE} = I_B r'_{bb} + V'_{BE} \quad (10.1)$$

$$= \frac{I_C}{h_{FE}} r'_{bb} + V'_{BE} \quad (10.2)$$

As indicated by this equation, the base-to-emitter voltage diminishes with a rise in temperature for low values of collector current, but tends to increase with a rise in temperature for higher values of collector current.

Collector-to-Emitter Saturation Voltage

The collector-to-emitter saturation voltage $V_{CE,sat}$ is affected primarily by collector resistivity γ_C and the amount by which the natural gain of the device h_{FE} exceeds the gain with which the circuit drives the device into saturation. This latter gain is known as the *forced gain* h_{FEf}.

At lower collector current, the natural h_{FE} of a transistor increases with temperature, and the *IR* drop in the transistor is small. The collector-to-emitter saturation voltage, therefore, diminishes with increasing temperature if the circuit continues to maintain the same forced gain. At higher collector currents, however, the *IR* drop increases, and gain may decrease. This decrease in gain causes the collector-to-emitter saturation voltage to increase and possibly to exceed the room temperature (25°C) value. Figure 10-10 shows the effect of temperature of the collector-to-emitter saturation voltage.

Collector Leakage Currents

Reverse collector current I_R is a resultant of three components, as shown by the following equation.

$$I_R = I_D + I_G + I_S \tag{10.3}$$

Figure 10-11 shows the variations of these components with temperature.

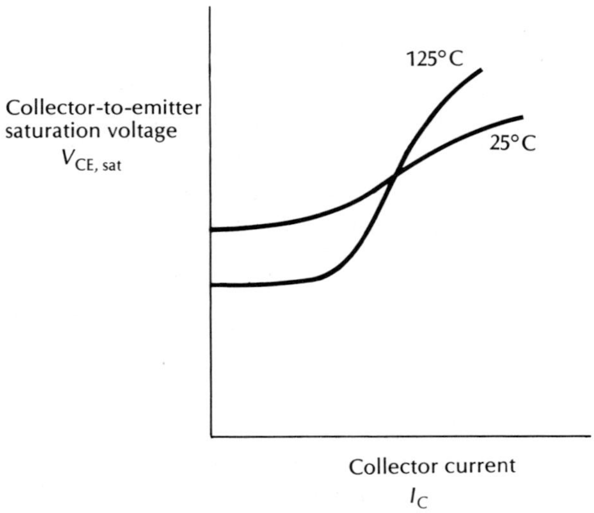

FIGURE 10-10 Collector current as a function of collector-to-emitter saturation voltage at different temperatures.

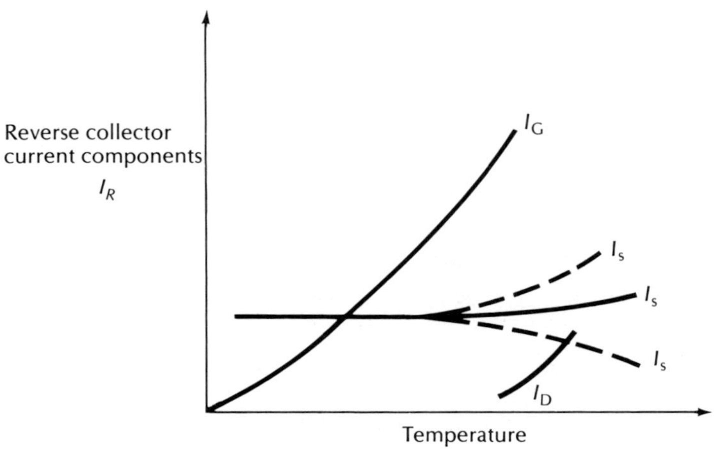

FIGURE 10-11 Reverse collector current as a function of temperature.

The diffusion or saturation current I_D is a result of carriers that diffuse to the collector–base junction and are accelerated across the depletion region. This component is small until temperatures near 175°C are reached. The component I_G results from charge-generated carriers that are created by the flow of diffusion carriers across the depletion region. This component increases rapidly with temperature. Currents I_D and I_G are referred to as *bulk leakages.* The term I_S represents surface leakage that is caused by local inversion, channeling, ions, and moisture. This leakage component is dependent on many factors, and its variations with changes in temperature are difficult to predict.

At low temperatures, either surface or bulk leakage can be the dominant leakage factor, particularly in transistors that employ a mesa structure. At high temperatures, charge generated carriers and diffusion current are the major causes of leakage in both mesa and planar transistor structures; the current I_G, therefore, is the dominant leakage component. Because of the dominance of surface leakage I_S at low temperatures and the fact that this leakage may vary either directly or inversely with temperature, it is not possible to define a constant ratio of the leakage current at low temperatures to that at high temperatures. In view of the fact that power transistors are normally operated at high junction temperatures, it is more meaningful to compare the leakage characteristics of both mesa and planar transistors at high temperatures. The relative reliability of different types of power transistors, which is in no way related to the magnitude of low temperature leakage current, is also best compared at high temperatures.

Leakage currents are important because they affect biasing in amplifier applications and represent the "off" condition for transistors used in switching applications. The symbol I_R used in the preceding discussion represents any of several different leakage currents commonly specified by transistor manufacturers. The most basic specification is I_{CBO}, which indicates the leakage from collector to base with the emitter open. This leakage is simply the reverse current of the collector-to-base diode. In addition to the I_{CBO} value, I_{CEV}, I_{CEO}, and I_{CER} specifications are often given for transistors. Current I_{CEV} is the leakage from the collector to emitter with the base–emitter junction reverse-biased. Current I_{CER} is the leakage current from the collector to the emitter with the base and emitter connected by a specified resistance. Current I_{CEO} is the leakage current from collector to emitter with the base open. Current I_{CEV} differs from I_{CBO} only very slightly, and in most transistors the two parameters can be considered equal. (This equality is not maintained in symmetrical transistors.) Current I_{CEO} is simply the product of I_{CBO} at the voltage specified and the h_{FE} of the transistor at a base current equal to I_{CBO}. Current I_{CEO} is, of course, the largest leakage current normally specified. Current, I_{CER} is intermediate in value between I_{CEV} and I_{CEO}.

Second Breakdown in Bipolar Power Transistors

Second breakdown is a potentially destructive phenomenon that occurs in all bipolar transistors. It results when the energy absorbed by the transistor exceeds a critical level and causes localized hot spots within the transistor pellet. The start of second breakdown is characterized by an abrupt decrease in collector-to-emitter voltage with a small dynamic resistance in the second breakdown region. Although second breakdown may result from several modes of transistor operation, this phenomenon can be broadly categorized into two major classes: (1) forward-biased emitter-to-base second breakdown, which occurs when the transistor operates in the active region; and (2) reverse-biased emitter-to-base second breakdown, which occurs during the cutoff mode of transistor operation.

Forward-Biased Second Breakdown

When the transistor is heavily forward-biased into the active region, a transverse electric field is produced in the base region, and a space-charge layer is formed at the base–collector junction. As current flows from emitter to collector, the transverse field focuses the current into a narrow region below the emitter edge. When the current flows through the space-charge layer, a significant amount of heat is generated. With the current focused into a small area, the heating effect is localized in this area, and hot spots may be formed within the silicon pellet. If unchecked, these hot spots initiate a regenerative cycle of high-density current that results in forward-biased second breakdown in the transistor.

The carrier concentration in the pellet and, consequently, the severity of the hot spots are determined mainly by the magnitude of the transverse base field and by the applied collector voltage, which determines the intensity of the electric field across the space-charge layer formed at the base–collector junction. The magnitude of the transverse base field depends on the width of the transistor base, the base resistance, and the spreading of the space-charge layer that results from the application of collector voltage. The transverse base field increases with increases in base current that result from higher injection levels and reduced current gain at the higher injection levels.

The severity of the hot spots is inversely proportional to the width of the transistor base and directly proportional to the magnitude of the applied collector voltage. As a result, the current level $I_{s/b}$ at which second breakdown occurs decreases rapidly with an increase in applied collector voltage. The $I_{s/b}$ capability is also decreased as the transistor frequency capability f_T is increased. The thermal capacitance of the transistor pellet results in a localized thermal time constant that restricts instantaneous formation of hot spots. Selecting a transistor that has the lowest frequency capability, consistent with circuit requirements, minimizing high voltage transients as much as possible on the supply line, and minimizing the on time of the transistor will all serve to maximize the $I_{s/b}$ capability of the transistor.

Reverse-Biased Second Breakdown

When current flows through a power transistor in which the emitter-to-base junction is reverse-biased, the direction of the transverse base field is opposite that produced in a forward-biased transistor. As a result, the emitter current is focused into a small region at or near the center of the emitter. Because the current is crowded into a smaller region under reverse-bias conditions, reverse-bias second breakdown can be encountered at substantially lower energy levels than those at that forward-bias second breakdown occurs.

The resistance of a transistor to reverse-bias second breakdown is reduced by any design alteration that increases current density or prevents spreading of emitter current. In power transistors that have a narrow base, an accelerating base field, or insufficient emitter size for their operating current, second breakdown generally occurs at lower energy levels than in transistors without these factors.

Reverse-bias second breakdown is usually described in terms of energy because voltage, current, and pulse duration are interdependent when the emitter–base junction is reverse-biased. The transverse base field in the transistor depends, to a large extent, on the turn-off base current and turn-off voltage. It is natural, therefore, to assume that the energy level at which second breakdown takes place is a strong function of the input-circuit turn-off voltage V_{BE} and the series resistance R_{BE}. If R_{BE} is increased and V_{BE} is decreased, the turn-off base current and the transverse base field are both reduced, and the second breakdown energy level is raised. Because turn-off time is an inverse function of base currents, the circuit designer must compromise between transistor turn-off speed and second-breakdown considerations.

Bipolar Transistor Drive Circuits

Transistors used as switches are intended to quickly turn on, provide as little forward drop as possible while conducting current, and then quickly turn off. Ideally the only losses incurred would be the power dissipated while the transistor is on, with no loss during the turn-on or turn-off time. However, as described earlier, there are delays in switching, and these delays play an important role in determining the total power loss in the transistor during a switching period. How the transistor base is driven will impact its performance. The bipolar transistor is a current amplifying device. If a particular device has a gain (h_{FE}) of 10 and is to pass 10 A, then a base drive of 1 A will be required to ensure full saturation. Insufficient base current will result in less than full turn-on, with a greater voltage drop from collector to emitter than desired. This higher voltage will cause unnecessary power loss and heating within the transistor. The circuits described in the following paragraphs are offered as some of the possibilities for bipolar transistor base drive.

The most common method of achieving a higher gain in the power device is to use two or more transistors in a Darlington arrangement (Fig. 10-12). As shown in Fig. 10-12, base current for Q_2 is supplied from the emitter of Q_1. The total gain of the combination is then increased. If each transistor has a

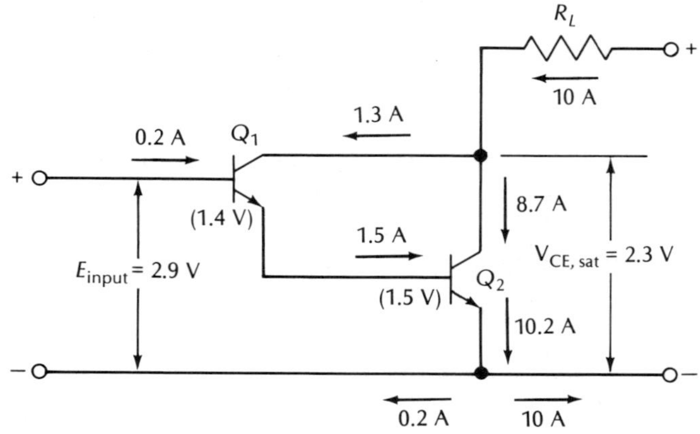

FIGURE 10-12 Darlington configuration for increased gain.

gain of 10, at the current it will be passing, then the total gain for the Darlington will be 100. Only 0.1 A of base current into Q_1 will be needed to allow 10 A of collector current in Q_2. A drawback to the Darlington arrangement is higher $V_{CE,sat}$. The total saturation voltage of Q_2 will be the sum of the $V_{CE,sat}$, Q_1 and the base to emitter voltage, $V_{BE,sat}$, of Q_2. The power dissipated in this arrangement will then be the load current times the $V_{CE,sat}$ of Q_2. This is only the saturation loss, though, and the switching losses must still be taken into consideration. This power transistor arrangement is operating at a forced gain of 100, so the base current required to operate a 10 A load is only 0.1 A. But suppose it is to operate from a CMOS control circuit where much lower currents are available. One solution is to keep adding additional transistors in the Darlington configuration, with their gains multiplying, until the desired gain is achieved. But the cost in total $V_{CE,sat}$ will cause unnecessary power loss. An alternative approach is shown in Fig. 10-3. The total forced gain can now be extended by the addition of one more transistor, without the additional increase in power loss in the output transistor. In this arrangement the saturation voltage of Q_3 is the same as it was in Fig. 10-12, and the total load current will be the sum of the collector currents.

A similar darlington arrangement can be utilized in an emitter–follower stage operating from a single input, as shown in Fig. 10-14. This arrangement is commonly found as the pass stage in switching regulators. In this arrangement $V_{CE,sat,Q_2} = V_{BE,sat,Q_1} + V_{BE,sat,Q_2} + V_{R_1}$. The value of R_1 must be kept small in order to minimize the drop across, and ultimately the loss within, the pass stage. As the value of R_1 is decreased, however, the current that the control circuit must handle increases. Figure 10-15 uses an arrangement similar to that shown in Fig. 10-13 to help solve this problem. In this arrangement the addition of a third transistor reduces the current requirement through R_1, due to the additional gain of the circuit. Now $V_{CE,sat,Q_3} = V_{BE,sat,Q_1} + V_{R_1} + V_{R_2}$. The control circuit will still handle the same current as before, but the total saturation voltage across the power stage will be reduced.

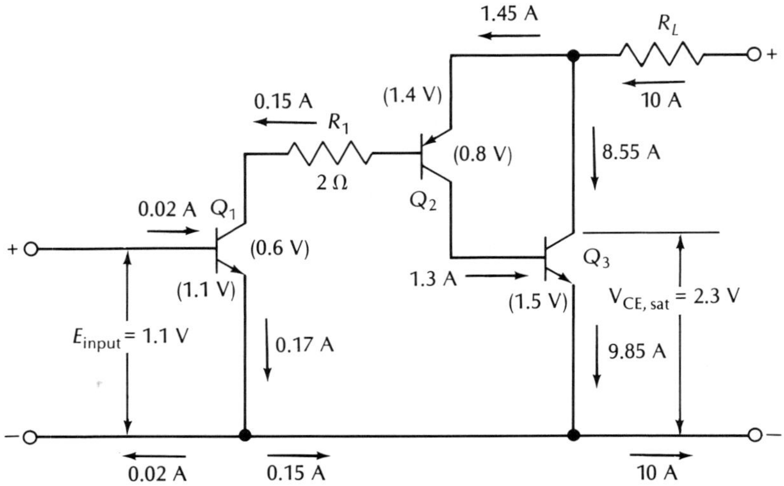

FIGURE 10-13 Higher gain over Fig. 10-12 without increased $V_{CE,sat}$.

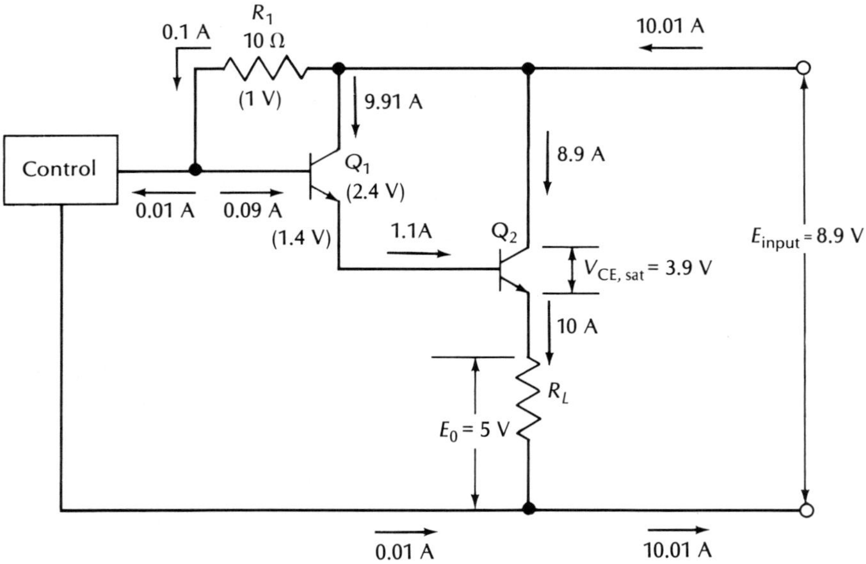

FIGURE 10-14 Darlington emitter–follower.

In the circuits described above, all the transistors are turned on at one time. Transistor Q_1 turns on and, in turn, turns on Q_2, and so on. Some arrangements use one transistor turning on to turn off another. Figure 10-16 illustrates such an arrangement. When Q_1 turns off, C_1 charges to V_{C,Q_1}. When Q_1 turns on, C_1 discharges through Q_1, R_4, Q_2 base–emitter, and R_3 to reverse bias Q_2. It is important for switching speed to ensure that the base of the

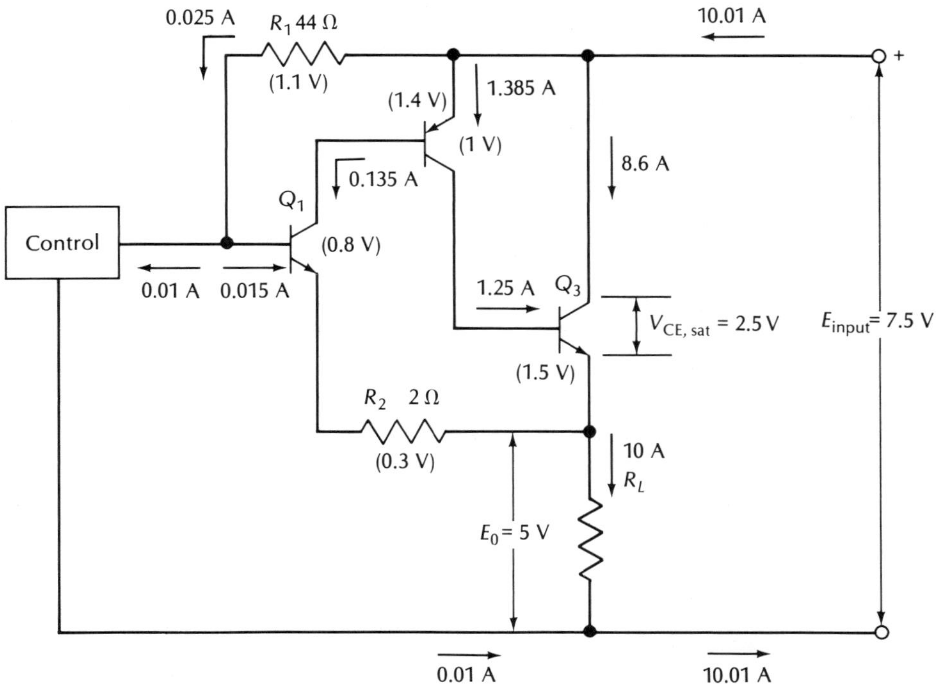

FIGURE 10-15 Emitter–follower with improved $V_{CE,sat}$.

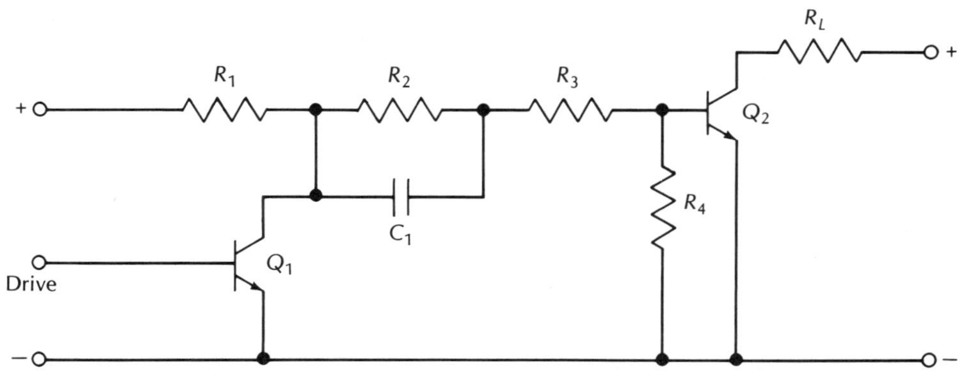

FIGURE 10-16 Direct-coupled base drive with reverse bias supplied by capacitor.

power transistor is reverse-biased, if possible. An advantage of this circuit is the addition of C_1 provides current flow around R_2 for current peaking during the turn-on and turn-off periods. During the steady state conduction period of Q_2, C_1 is out of the picture and R_2 joins in to limit the steady state base current in Q_2. A disadvantage of this circuit is that Q_1 must conduct all the current through R_1, during the time that Q_2 is to be off, as well as the reverse

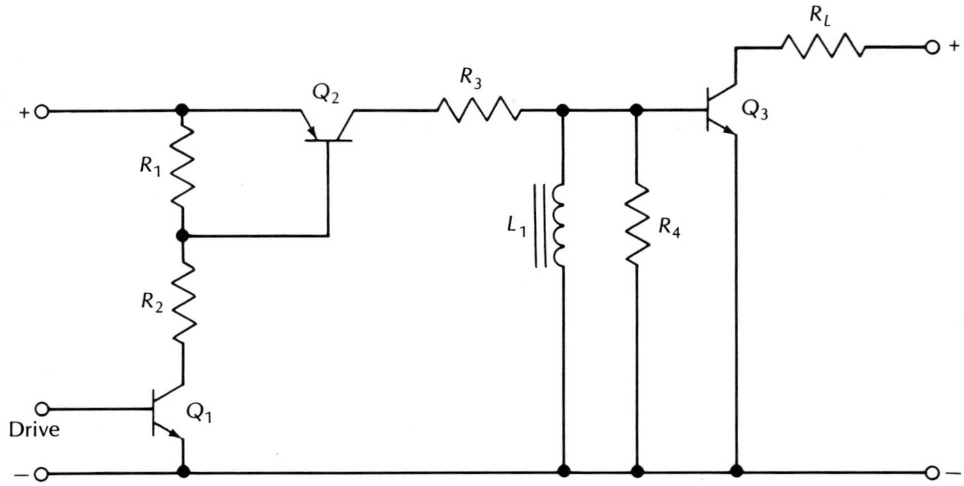

FIGURE 10-17 Direct-coupled base drive with reverse bias supplied by inductor.

base current of Q_2 at the moment of turn-off. Another circuit that works well to provide reverse bias during turn-off is shown in Fig. 10-17. In this circuit all the transistors turn on and off together. When Q_1 turns on, it turns on Q_2, which then provides base current through R_3 to Q_3. Inductor L_1 will support the base voltage at Q_3 for a time, then begin to conduct current, reducing the current into the base of Q_3. Inductor L_1 stores energy so that when Q_1 turns off, interrupting current flow in L_1, a voltage reversal across L_1 will reverse bias the base of Q_3 causing a very fast turn-off. This type of drive circuit is most suitable for high frequency, low-duty-cycle applications.

Figure 10-18 shows another direct coupled drive circuit that provides fast switching. In this circuit a pair of transistors, Q_1 and Q_2, operate from the same control signal to turn Q_3 on and off. When Q_1 turns on, C_1 provides a fast current peak to get Q_3 turned on quickly. Resistor R_1 then limits the steady state current for the remainder of the cycle. During the time Q_3 is on, C_1 is charged positive. When Q_1 turns off, Q_2 turns on and instantly discharges C_1, which provides reverse bias to the base of Q_3, allowing Q_3 to quickly turn off. In some instances inductive load currents will cause high voltage spiking on the collector of Q_3 when it switches off. This is caused by the switching speed of Q_3, which we are trying to minimize to reduce losses. In cases this high spiking is not present on every high-frequency cycle. This is the case, for instance, where Q_3 is switching one end of a push–pull power transformer with sine-weighted pulse-width modulation (PWM), with the ultimate output voltage waveform to be a sine wave. Proper care in the design of the power transformer, as well as other circuit characteristics, will dictate that the highest voltage spikes will appear only during the widest pulse widths at Q_3. If these spikes become too high, the voltage rating of Q_3 will be exceeded and Q_3 will fail. Typically a dissipative snubber circuit in parallel with Q_3 is a cheap solution to keeping the voltage spike under control. But that snubber will be working on every PWM pulse, and dissipating power with each. An alternative to the snubber is shown in Fig. 10-18 with the addition of a zener diode. In this case the zener is not expected to dissipate the energy of the voltage spike, but rather it will conduct when the spike reaches the zener voltage, providing current from the voltage spike back into the base of Q_3, turning Q_3 back on. So long as the voltage spike remains high enough to break over the zener, Q_3 will stay on and clamp its own collector voltage to a safe level. Now, instead of having power dissipated on every pulse in a snubber circuit, Q3 dissipates the power itself, but only on pulses that cause a sufficiently high spike voltage to be concerned. The resistor in the base of Q_2 is there to allow the current returned to the base of Q_3 from the zener to turn off Q_2 long enough to allow Q_3 to snub its own voltage spike. As soon as the voltage is reduced to a safe level, the zener stops conducting and Q_3 turns back off.

Transformer-coupled drives are often used to provide isolation where necessary. Examples of two transformer-isolated drive circuits are shown in Fig. 10-19. Figure 10-19*a* uses a push–pull primary. Transistors Q_1 and Q_2 operate out of phase with each other. Starting with Q_1 on, the voltage induced on the secondary is polarized negative for base drive at Q_3, so Q_3 is off. When Q_1 turns off and Q_2 turns on, the voltage on the secondary of T_1 reverses and turns on Q_3. The diode connected between T_1 and Q_3 collector provides a Baker clamp. It provides decreased storage time by maintaining the base–collector junction in reverse bias and not allowing Q_3 to go into deep saturation. When Q_3 turns on, its base capacitor will be charged to the voltage drop of the two series diodes, approximately 2 V. When Q_2 turns off and Q_1 turns on, the secondary voltage of T_1 is again reversed to turn off Q_3. The reversal

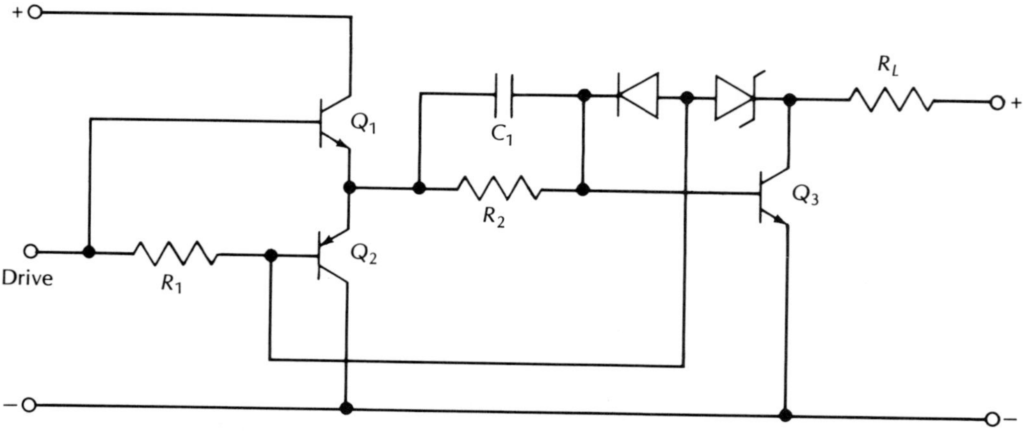

FIGURE 10-18 Direct-coupled drive with zener diode protection.

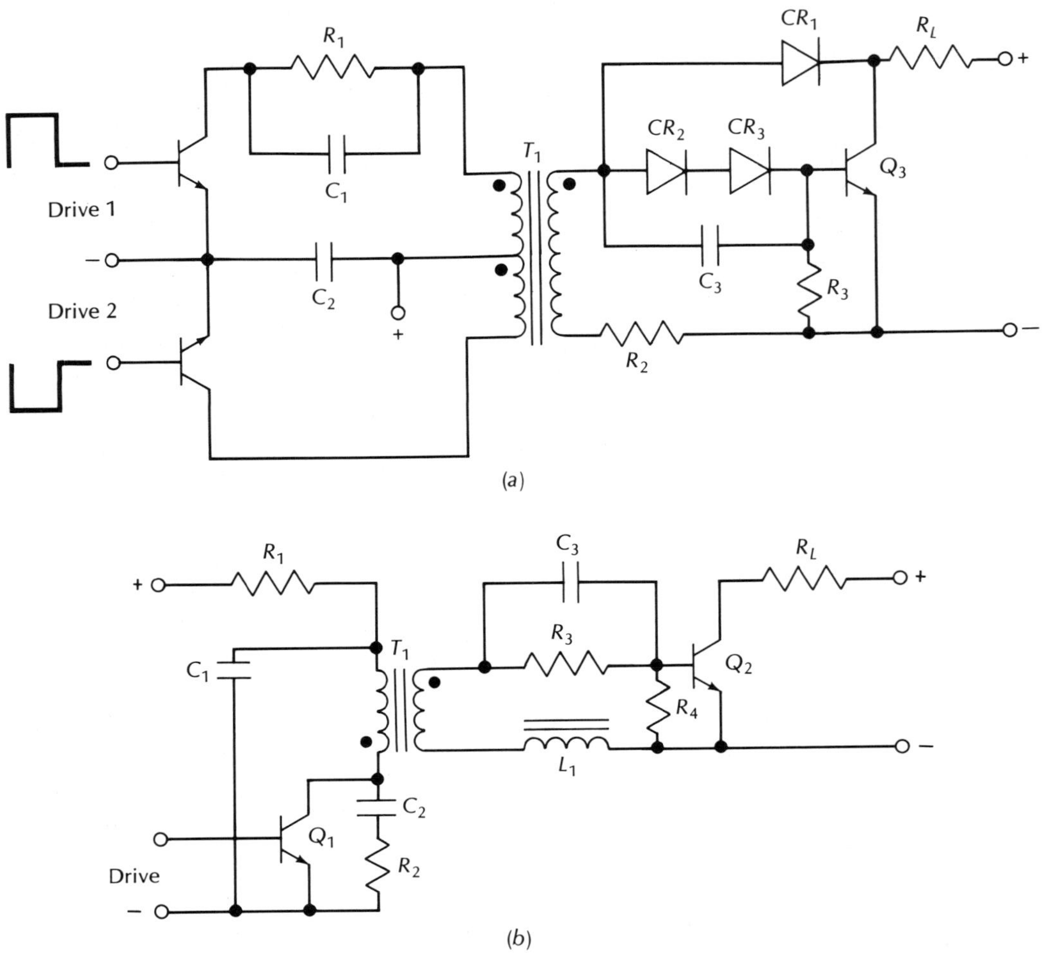

FIGURE 10-19 Transformer-isolated base drives.

of the secondary voltage causes the base capacitor to quickly discharge, imposing a reverse bias on Q_3 and providing a fast turn-off. This circuit works well for 50% duty-cycle applications.

In Fig. 10-19*b* a flyback drive transformer is used in place of the push–pull approach. While Q_1 is on, the current is limited by R_1. When Q_1 turns off, the energy stored in the primary of T_1 is transferred to the secondary to turn on Q_2. Current flows through R_3, charging C_3, Q_2 base–emitter, and L_1. The inductance of L_1 is normally set equal to the leakage inductance of T_1. When Q_1 turns off, C_1 charges through R_1. When Q_1 turns on again, C_1 discharges across T_1 and the secondary voltage reverses to reverse-bias Q_2 for turn-off. R_3 can be replaced with diodes as shown in Fig. 10-19*a* to prevent C_3 from discharging through R_3, extending the reverse bias on Q_3.

All the circuits discussed thus far require constant current from the control circuit during either the on or off time of the final output transistor. In cases where very high output collector currents are to be controlled, and a Darlington approach is not desired, there is another method that could be used. Figure 10-20 shows one example of a proportional base drive circuit. It is called a "proportional drive" because of the arrangement of the windings on the isolation transformer. Once collector current is established, the current flowing through winding N_4 to the base of Q_2 is proportional to the current flowing through N_3 to the collector of Q_2. As more collector current is demanded by the load, more base current will flow to support the demand. This is independent of the primary circuit, so during the on time of Q_2, no current is required from the control circuit, other than what is required to get Q_2 turned on or off. When Q_1 is on, Q_2 is reverse biased as the current increases exponentially through $T_1 - N_1$. Resistor R_1 limits the peak current. When Q_1 turns off, the energy in N_1 is transferred to N_4, which forward-biases Q_2. Before the base current to Q_2 can start to decay, collector current starts to flow through N_2. While Q_1 is off, C_1 charges through R_1. When Q_1 turns on again, N_3 is initially short-circuited. Then C_1 discharges through N_1 and all windings reverse in polarity. Transistor Q_2 is then reverse biased and turns off. A major advantage of this type of drive circuit is the ability to control very large load currents with very small amounts of control current, without the addition of Darlington transistors and their relatively large voltage drop. Fairly low saturation voltages at Q_2 can be realized, along with relatively fast switching time. The major disadvantage of this circuit is the requirement for the drive transformer to be constructed with a winding that must handle the full load current.

The introduction of power MOSFET devices has not eliminated the use of bipolar power transistors in switching converters. In fact, it has caused the

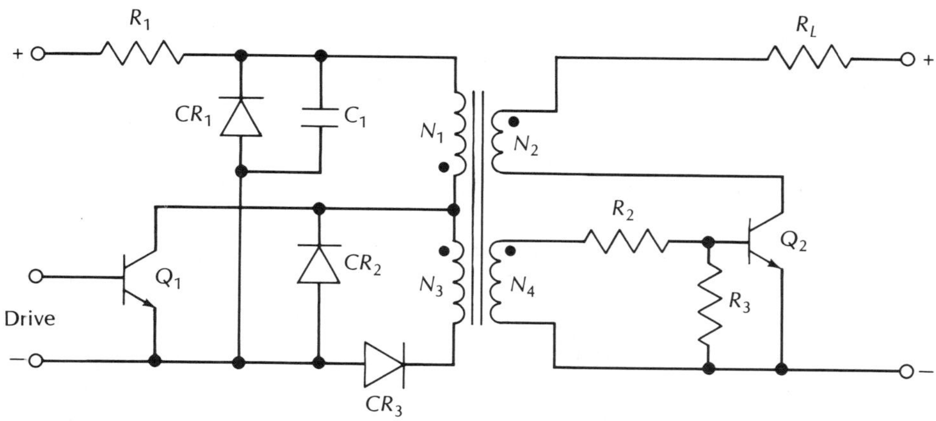

FIGURE 10-20 Transformer-isolated proportional base drive circuit.

bipolar manufacturers to advance their own technology to the point of making very fast transistors that provide high voltage capability, low $V_{CE,sat}$, and low cost. In higher power converters, operating below 40 kHz, the bipolar transistor still has some advantages over power MOSFETs and should be given consideration.

MOSFET Considerations

If the MOSFET is being considered as the switch for your design its recommended that you first read the preceding discussion on bipolar transistor characteristics, if for no other reason than to allow you to fully appreciate how much simpler the MOSFET is to implement. Being a voltage controlled device, the MOSFET gate drive circuitry can be much simpler than the bipolar base drive circuits, and power dissipation in the drive circuit can be greatly reduced. Switching time is independent of temperature, and storage time delay is eliminated. The safe operating area extends over the entire current and voltage rating of the device, so second breakdown failure, as experienced in the bipolar device, has been eliminated.

Overall, the cost and performance of power MOSFET transistors have nearly eliminated the use of bipolar devices in modern switching converters. Some applications still find bipolar transistors are the best choice, but MOSFET technology advances are continuing to narrow down the bipolar niche. This is not to imply that MOSFET technology is without problems as well. There are many considerations that must be taken into account when applying MOSFET transistors to achieve success.

MOSFET Switching Characteristics

In order to turn on a MOSFET, a voltage must be applied between the gate and source terminals. This will produce a flow of current in the drain. The gate is electrically isolated from the source by a layer of silicon oxide. Although in theory no current flows into the gate when DC voltage is applied, in practice a small leakage current, in the order of nanoamperes, will flow into the gate. Prior to the application of voltage gate-to-source, the impedance between the drain and source terminals is very high, and only a small leakage current flows in the drain until the applied voltage exceeds the drain-to-source avalanche voltage. This can be demonstrated with any MOSFET and an analog ohmmeter. First short-circuit the gate and source terminals to ensure no charge is stored in the gate capacitance. Place the ohmmeter leads across the drain and source terminals, and note that in one direction you can measure the body diode within the MOSFET while in the other direction it displays very high impedance. Now, leaving the meter lead connected to the source, move the other meter lead to the gate terminal, forward biasing the gate-to-source terminals. Now by moving the lead from the gate back to the drain you will find a very low drain-to-source impedance. The MOSFET has been left in an on condition due to the stored charge on the gate capacitance.

The voltage applied between the gate and source terminals causes an electric field to be set up within the device. This field modulates the resistance between the drain and source terminals and permits a current to flow in the drain in response to the applied drain circuit voltage. The amount of current that flows depends on the amount of voltage applied to the gate, assuming that the impedance of the external drain circuit is not limiting. Because only leakage current is drawn by the gate the DC current gain is extremely high, typically in the order of 10^9. Current gain is a rather meaningless parameter for a power MOSFET, and it is seldom used. Because a flow of current in the drain is produced by a voltage applied to the gate, a more useful parameter is transconductance. This is the change of drain current brought about by a 1-V change of gate voltage.

Metal oxide semiconductor field-effect transistor power transistors are much faster than bipolar power transistors of comparable size, primarily because they do not have minority carrier delay times. The response times of MOSFETs are determined primarily by the device capacitances, and secondarily by such factors as the extremely short channel transit time of the electrons. The input capacitance C_{iss}, in its two components (C_{gs} and C_{gd} or C_{rss}), is the primary factor that determines the response time of a MOSFET. Figure 10-21 shows a typical switching waveform when driven by a 500-ns wide pulse. A good first-order understanding of the switching response times $t_{d(on)}$, t_r, $t_{d(off)}$, and t_f, can be made by considering the power MOSFET an ideal switch with a 3-V "on" threshold. The output current can be considered controlled, without delay, by the instantaneous gate voltage, which, in turn, is controlled by the time constant formed by $R_{input} \times C_{iss}$.

The turn-on delay $t_{d(on)}$ is primarily the time for the transistor input capacitance to be charged by the control signal to the threshold level of 3 V maximum. The rise time t_r is the gate charging time required to drive from the threshold voltage, through the linear control region, to the gate voltage (typically 5–8 V) required for full conduction of the drain current. In turn-off the procedure is reversed, and the turn-off delay time $t_{d(off)}$ is primarily the time required for the gate capacitance to discharge from the gate overdrive saturation voltage (typically 10 V) to the active gate control region. Finally, the fall time t_f is primarily the time delay required for the input capacitance to discharge through the active control region to the gate threshold voltage. This demonstrates the importance for having a low-impedance gate drive circuit since the higher the gate drive resistance, the longer the switching times will be.

The switching times described above apply for a resistive load. In many applications the load during the switching interval will be essentially inductive, and it is often of more practical interest to the circuit designer to know

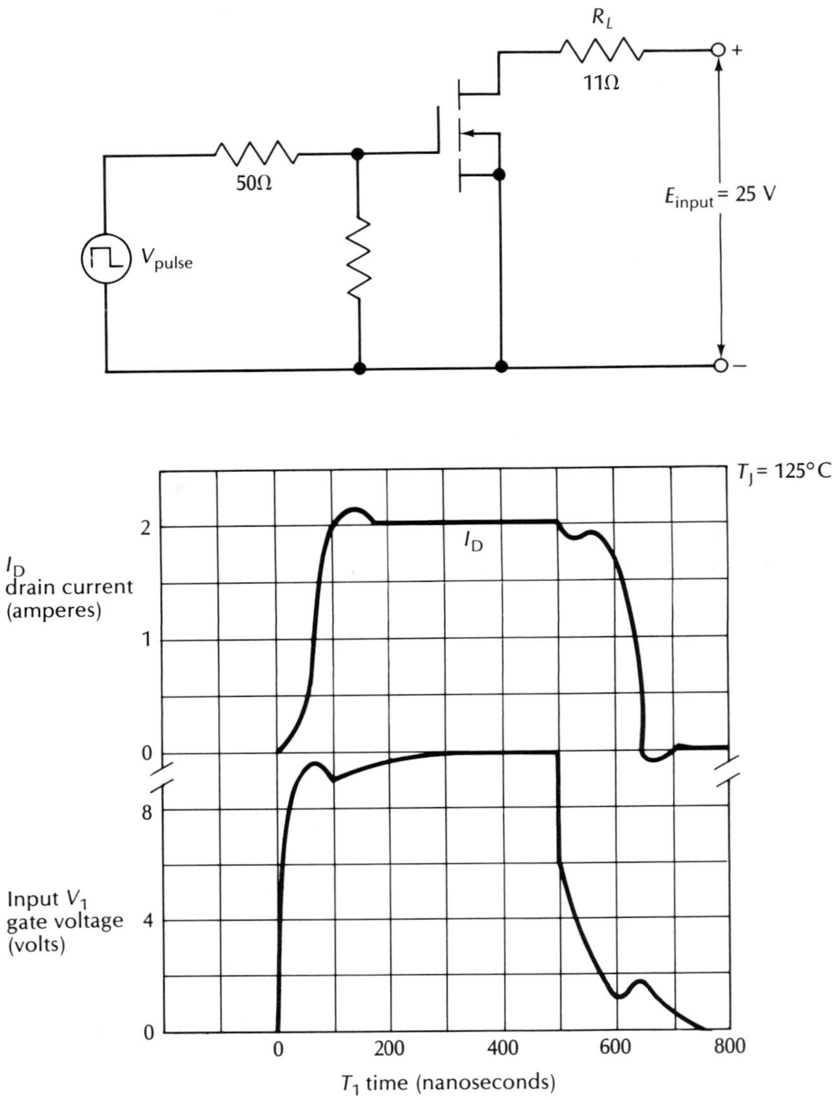

FIGURE 10-21 Typical switching waveform of an IRF-330.

the switching times with an inductive load. Figures 10-22 and 10-23 show a simple circuit with a clamped inductive load and the idealized switching waveforms. Note that at turn-on the current first rises, then the voltage falls. At turn-off the drain voltage rises, then the current falls. Note that this assumes that the load is sufficiently inductive that current flows continuously in it; through the MOSFET when it is turned on, and through the clamping diode when the MOSFET is turned off. It is also assumed that stray circuit inductance between the clamping diode and the MOSFET is negligible. Thus at turn-on, load current that is already flowing in the clamping diode com-

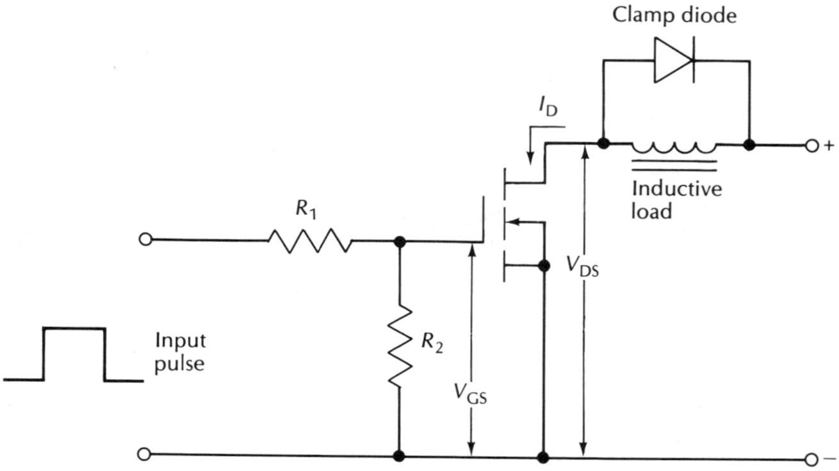

FIGURE 10-22 Metal oxide semiconductor field-effect transistor (MOSFET) circuit with clamped inductive load.

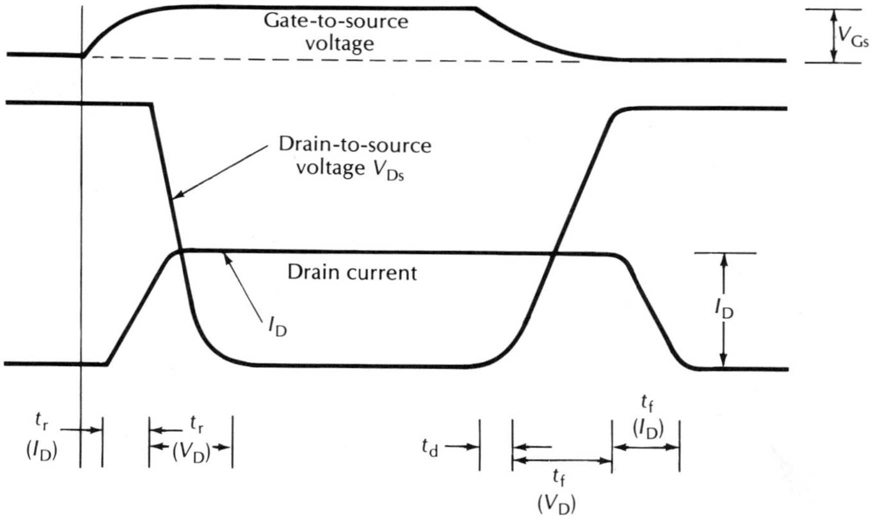

FIGURE 10-23 Idealized switching waveforms for clamped inductive load.

mutates into the MOSFET, and the transfer of this current is unrestricted by stray circuit inductance. In general, this assumption for the turn-on condition is somewhat pessimistic, but it can be approximated in certain circuits.

Temperature Stability in MOSFET Transistors

Temperature variations have little effect on MOSFET gain and switching time characteristics, relative to the temperature stability of bipolar transistors. The transconductance of MOSFET transistors typically varies less then 20% from the 25°C value, over a −55 to 125°C range. A rough comparison of the typical DC gain stability for a MOSFET versus a power bipolar would yield a transconductance g_{fs} temperature coefficient of about −0.2%/°C versus a bipolar current gain h_{FE} temperature coefficient of about +0.8%/°C. Since the threshold voltage (bias point) and the transconductance (DC gain) move with temperature in compensating directions, an operating point on the DC transfer characteristic maintains unusually good open-loop stability.

The switching time of a MOSFET is independent of operating temperature because the response times are dependent primarily on the input capacitances, which are essentially temperature invariant. The on- resistance $R_{DS(on)}$ of power MOSFETs has a positive temperature coefficient of 0.4–0.8%/°C, depending on voltage rating. This is an advantage in paralleling MOSFETs and also accounts for their excellent safe operating area. However, in determining the on-state power losses in a switching mode, the increased value of $R_{DS(on)}$ at the actual maximum junction operating temperature must be used. Sufficient heat sink must always be used so that a thermal runaway situation is avoided.

MOSFET Safe Operating Area

As discussed earlier, in bipolar transistors, second breakdown is a great concern and often the cause for failure in the device. MOSFETS do not display this second-breakdown phenomenon. If localized potentially destructive heating occurs within a MOSFET transistor, the carrier mobility in that area decreases. As a result, the MOSFET has a positive temperature coefficient and acts in a self-protective manner by forcing current to be uniformly distributed through the silicon die. The rated maximum power dissipation of a bipolar actually applies only over a very limited range of collector voltage, typically up to 10% of rated V_{CEO}. At higher voltages, the DC power dissipation is severely limited to second breakdown. Typically, at rated voltage the permissible power dissipation is only 5% of full-rated power. By comparison, the safe operating area for a MOSFET will allow the DC current to be limited through the voltage range only by the power dissipation slope. There are no secondary slopes in the DC dissipation curve that indicate the second breakdown limit commonly seen on bipolar safe operating area curves. The absence of second breakdown means that the power MOSFET is generally a much more rugged device than the bipolar transistor.

The V_{DS} absolute maximum rating of MOSFETs should not be exceeded by allowing them to operate in the avalanche region. However, it is possible to reliably turn off high-level inductive currents, which can generate high voltage inductive transients, by using a simple voltage clamp circuit. Because MOSFETs are very fast devices, caution must be taken that the voltage clamp device has a sufficiently fast response, and that it is closely coupled to the drain–source terminals. A zener diode connected physically as close as pos-

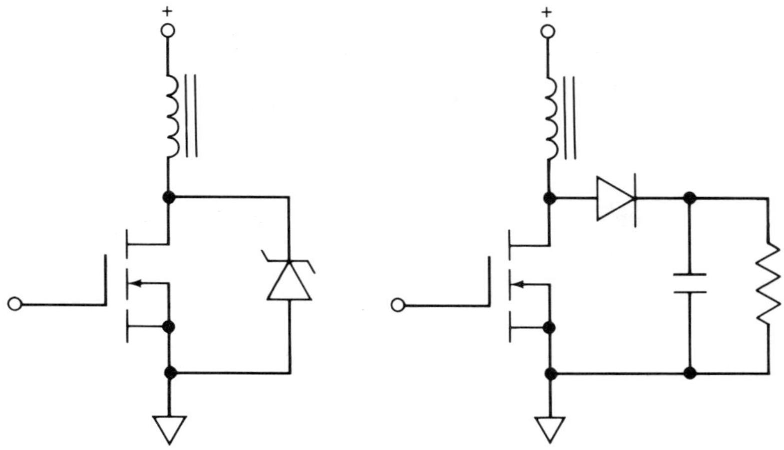

FIGURE 10-24 Example of voltage transient snubbing.

sible to the drain and source terminals generally will provide acceptable voltage clamping, as shown in Fig. 10-24*a*. Care must be taken, though, to ensure that the power dissipation in the zener does not exceed its ratings. The snubber circuit shown in Fig. 10-24*b* is more expensive, but it will also provide good transient suppression. The circuit is sized to sustain the constant voltage found across the drain to source during the switching cycle, but the capacitor will absorb energy only during transients and then dump the energy into the resistance during the remaining portion of the cycle. Since the need for the snubber stems from inductances in the circuit and the fast switching time of the MOSFET, it follows that the device(s) used to suppress the transient should be placed as close to the leads of the MOSFET as possible. There are cases where it is not possible because of circuit board layout and heat-sink location to get the snubber right next to the MOSFET. In these instances be sure to use a good wideband oscilloscope and measure the voltage transients appearing right at the drain to source of the part, not at the leads of the snubber circuit. The inductance contained in the short run from the MOSFET to the snubber is often enough to allow much higher voltages to appear at the leads of the MOSFET than what is seen at the snubber. Often times the best solution to a transient protection problem is to use a combination of both the snubber network and a zener. The zener is selected to clamp at a voltage slightly higher than the peak voltage allowed by the snubber circuit. That way the zener clamps when needed, but handles very little power.

Possibly one of the most overlooked problems associated with the application of power MOSFET devices in power converters is *dv/dt* failure. There are three conditions during which high *dv/dt* can occur, each with its own threshold that must be exceeded before problems are encountered. Static *dv/dt* occurs when the device is supposed to be off and a sudden voltage transient appears across the drain and source terminals. This voltage transient is coupled to the gate of the device by the built in gate-to-drain parasitic capacitance C_{rss}. If the gate is not held low relative to the source, through an impedance sufficiently low to prevent the gate-to-source voltage from rising above $V_{GS(th)}$, then spurious turn-on will occur. In this instance the effect of high *dv/dt* on the device will be dependent largely on the gate impedance and the application of the device in the circuit. Generally, if a transient is coupled to the gate in this manner, and causes the device to begin to turn on, the act of the device turning on will reduce the V_{DS}, thus reducing the *dv/dt*, causing the device the then turn back off. This illustrates the importance of insuring proper gate termination to promote good noise immunity and is a good reason to never operate a MOSFET in a drive circuit that allows the gate to become "open-circuited." The second *dv/dt* concern is that which occurs when switching inductive currents as discussed above. Inadequate transient suppression will allow the V_{DS} rating of the device to be exceeded, leading to device failure. An additional consideration for studying the drain-to-source voltage transient is to also consider that there is additional inductance within the MOSFET package itself that contributes to the overall problem. This internal inductance will generally be quite low, compared with other circuit inductances to be dealt with, but it is there nonetheless. MOSFET manufacturers usually provide a value for this inductance in the data sheets.

The third *dv/dt* consideration, and perhaps the least recognized cause of mysterious failures, is associated with the drain-to-source body diode, which is an integral part of the MOSFET device. This diode is actually a parasitic bipolar transistor, and, as discussed earlier, bipolar transistors are particularly susceptible to failure due to second breakdown. This problem occurs in half- and full-bridge converters. Refer to the circuit shown in Fig. 10-25*a*.
An inductive load is being pulse-width-modulated by transistors Q_1 and Q_2. The intrinsic body diodes, D_1 and D_2, are providing the required current conduction path after the MOSFETs are turned off. When current is flowing as shown by I_1, while Q_1 is turned on, Q_2 is off. When Q_1 is turned off, load current I_2 is commutated to D_2. If Q_1 is allowed to turn on again while D_2 is still conducting, Q_2 may suffer damage because its diode is conducting while it experiences a rapid rise in V_{DS}. Failure can occur with *dv/dt* as low as 1 V/ns, which is pretty easy to achieve with MOSFETs. There are no elegant solutions to this one. Slowing down the *dv/dt* with snubbers or slowing down the MOSFET turn-on will help. The circuit shown in Fig. 10-25*b* is about the most reliable way to prevent the problem. The additional parts cost and power loss of inserting more diodes, and their associated voltage drops, in the circuit are not desirable. But by diverting the I_2 current around the body diode the MOSFET is protected against *dv/dt* failure. Making the series diode a Schottky part, with its lower forward drop than standard diodes exhibit, will lessen the impact of its presence. The reverse diodes need to be fast or ultra-fast-recovery types.

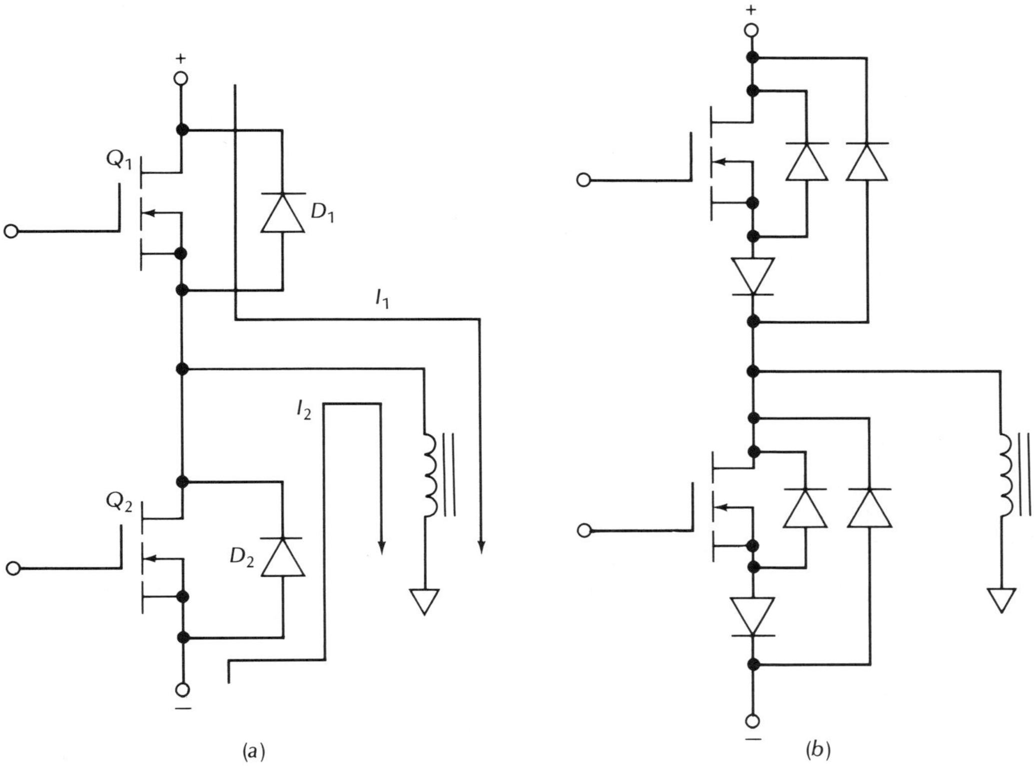

FIGURE 10-25 (*a*) Illustration of *dv/dt* problem in totem-pole circuit; (*b*) solution to *dv/dt* problem.

MOSFET Drive Circuits

The power MOSFET is a voltage-controlled device. By providing a positive voltage to the gate, with respect to the source, current will be made to flow in the drain. Reducing the voltage to zero will turn the drain current off. It seems simple enough, but several considerations must be taken into account, most of which concern switching speed. After all, that is the whole reason for considering the MOSFET as a power switch. Its $V_{DS,sat}$ compared to the $V_{CE,sat}$ of a similarly rated bipolar will often be higher, implying greater losses. In actual application, however, at high switching frequencies, a properly driven MOSFET will usually display lower losses, as reflected by its heating, than will a properly driven bipolar in the same application. This is due to the much faster switching capability of the MOSFET, resulting in lower switching loss. The following drive circuits are all intended to provide some guidance in providing proper gate drive, with emphasis placed on fast rise and fall times on the drive. During the on time of the device, the drive circuit must provide only the sustaining voltage to the gate, with negligible current flow. It is only during the initial turn-on of the gate that, as a result of the input capacitance

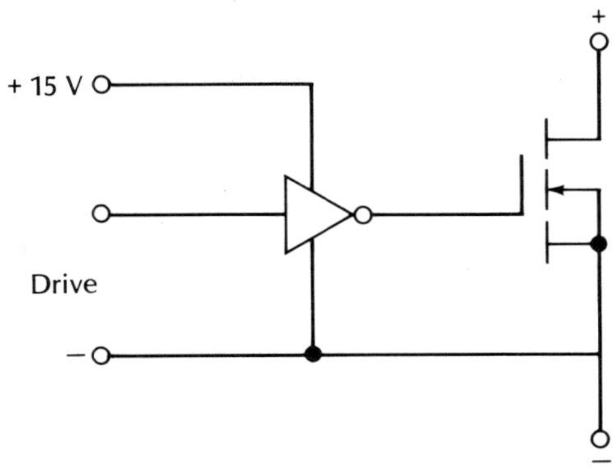

FIGURE 10-26 Simple direct-coupled gate drive using CMOS.

of the device, high current will flow, and that will often times be the limiting factor in switch speed.

Simplicity is of great importance in gate drive circuits since low cost usually follows. Figure 10-26 illustrates this point. In this drive circuit a single CMOS inverter gate, operating from a 12-V supply, will provide all the requirements to adequately turn on and sustain drain current in the MOSFET, all the way up to the maximum rating of the device. The inverter will both sink and source current, so it will also provide the requirement for holding the gate safely low during the time the device is off. Low-voltage, low-current MOSFETs, such as an IRFD121, have a low input capacitance. As the voltage and current ratings increase, the input capacitance specified also increases. Most gate drive circuits are accurately represented as a voltage source in series with some internal fixed resistance. Understanding that the MOSFET is controlled by gate charge helps in predicting the effect of the gate drive impedance on switching speeds. While The circuit shown in Fig. 10-26 may provide adequate current to switch a small MOSFET, with good switching times, a larger part will not switch as fast because of the current limitation of the single CMOS inverter trying to quickly charge the input capacitance. As the size of the part increases, it is important to keep this in mind and consider the source impedance of the drive circuit and how it will affect switching speed. Another example of a direct-coupled drive is shown in Fig. 10-27. This is simply an extension of the previous example, but all the CMOS gates within one package have been placed in parallel to provide more drive current capability. There are no special considerations or precautions to observe in doing this as the CMOS will parallel up very nicely. This is about as simple as it gets. From here, if more drive capability is required, it starts getting more expensive. If the CMOS inverter is not enough, then the next step may be to replace it with a similar device in function, but one that is designed to provide more drive current. Or you can build your own. Figure 10-28 shows a drive circuit made up of a CMOS logic gate driving a *PNP–NPN* transistor pair.

Properly driven, small bipolar transistors can switch very fast and provide the low source impedance required to quickly turn a power MOSFET on and off. A variation on this same theme is shown in Fig. 10-29, where two *NPN* bipolar transistors are driven in push–pull fashion to provide rapid turn-on and turn-off gate drive.

Isolated gate drives can be accomplished with both transformers, as illustrated in the section covering bipolar base drives, or with an optocoupler. Two problems arise with the use of an optocoupler. An auxiliary voltage supply must be created, operating on the output side of the optocoupler, and the optocoupler must be able to respond fast enough to ensure minimal delay between the control and the MOSFET turn-on and turn-off. Figure 10-30

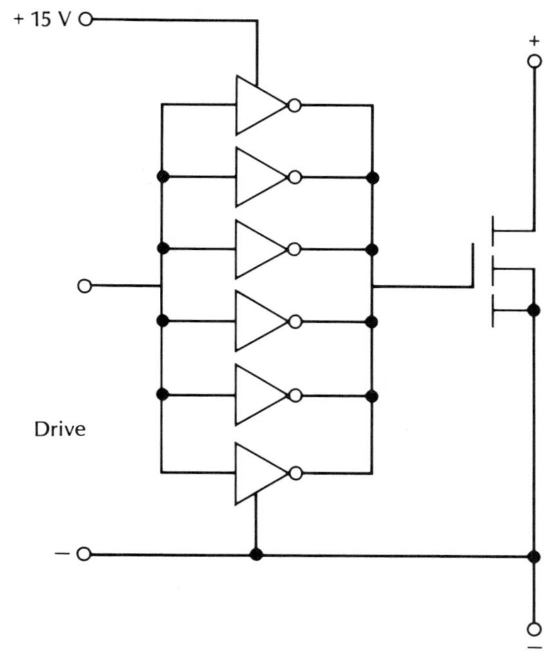

FIGURE 10-27 Paralleled CMOS to provide more drive capability.

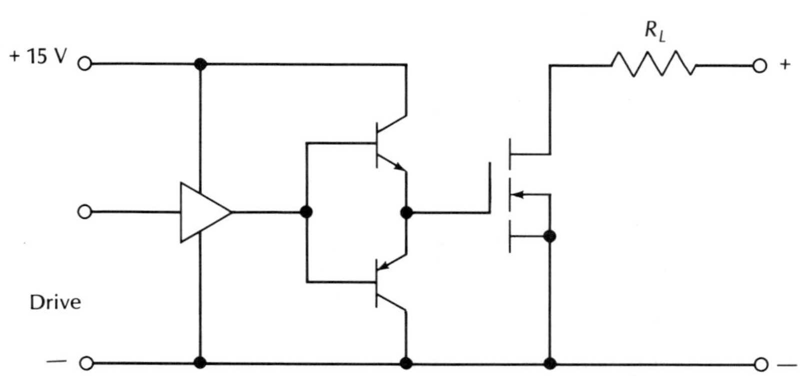

FIGURE 10-28 Direct-coupled gate drive using bipolar transistors to provide turn-on and turn-off current requirement.

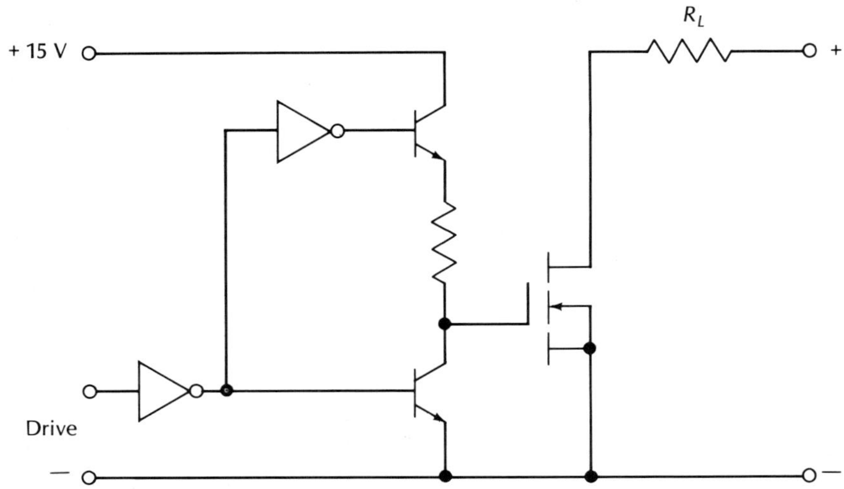

FIGURE 10-29 CMOS inverter gates provide push–pull drive to two *NPN* bipolar transistors.

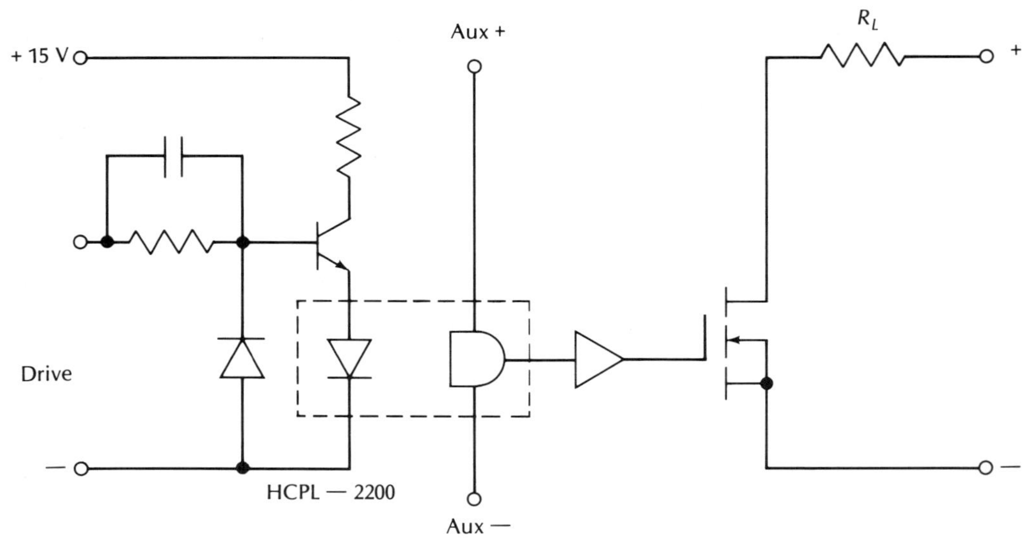

FIGURE 10-30 Isolated gate drive provided by fast optocoupler (Aux = auxiliary power supply).

shows an example of an optocoupler used for isolation to drive the gate of a power MOSFET. This particular optocoupler is quite fast and provides adequate speed to properly operate the MOSFET. The input to the optocoupler is switched by a small bipolar transistor, which can be operated directly from a low current source, such as a CMOS gate. The input base capacitor to the bipolar is necessary to allow the transistor to switch fast enough to be effective in this application. An auxiliary power supply provides the voltage on the output side of the optocoupler and powers the control circuitry between the optocoupler and the gate of the MOSFET. This particular circuit has been used successfully, but there are almost an unlimited number of variations on this theme that will work equally well. Think of the output of the optocoupler as just a signal source, and then apply any type of drive circuit between it and the MOSFET as required.

Like optocoupler drive circuits, transformer drive will also provide isolation. But the transformer has the ability to transfer power to the secondary, so the need for an auxiliary power supply is eliminated. Many of the transformer drive circuits described in the section on bipolar drive circuits will apply to MOSFETs. The main difference is the need for voltage rather than current. Typically the MOSFET requires 10 V gate-to-source to ensure full saturation. Aside from the initial high current pulse necessary to charge the input capacitance, no current is needed, so the secondary wire of the transformer can be quite small. Figure 10-31 uses a bridge configuration on the

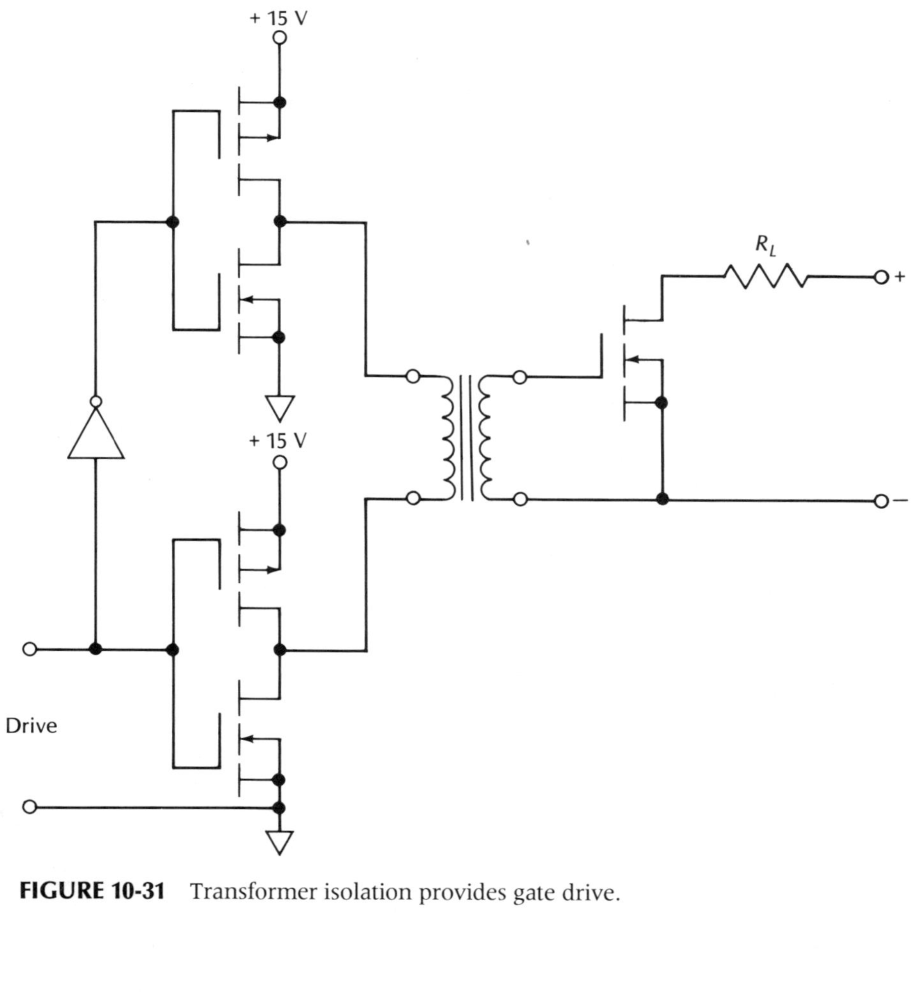

FIGURE 10-31 Transformer isolation provides gate drive.

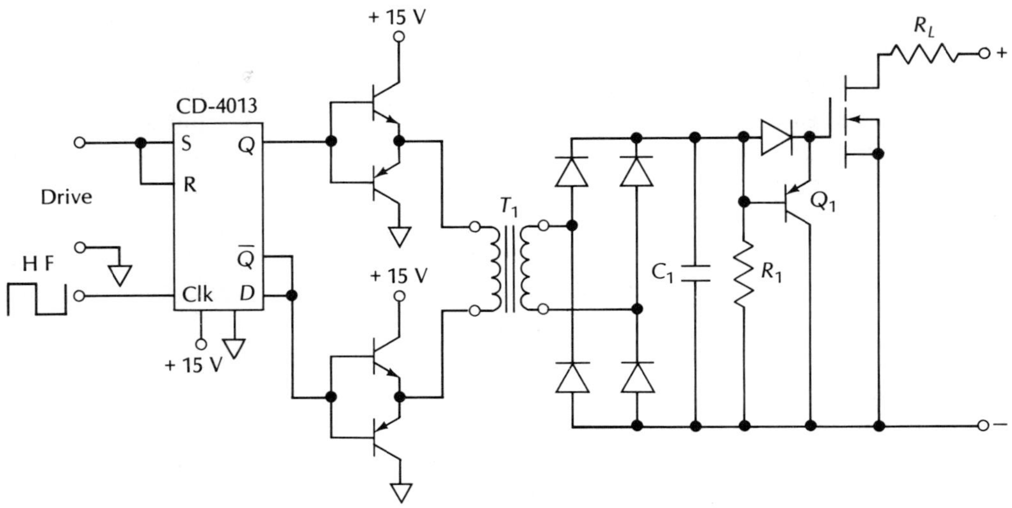

FIGURE 10-32 Transformer isolated gate drive utilizing high-frequency (2–10 MHz) clock to minimize the size of T_1.

primary of an isolation transformer. In this circuit small MOSFET devices have been used. The *P*-channel device conducts when its gate voltage goes negative with respect to the source. Each end of the primary is alternately switched between V_{CC} and common, at the switching frequency. The secondary of the transformer is wound to provide 10-V square-wave drive to the MOSFET. On alternate half-cycles the gate swings from +10 to −10 V, providing very fast turn-on and turn-off capability. In some instances it is not always necessary to provide PWM drive to all the power devices in the power stage. In the case of a full-bridge converter, for example, the designer may alternately switch the upper devices at a 60-Hz rate while providing PWM drive to the lower devices. If transformer isolation (as depicted in Fig. 10-31), is used for the high side drivers in this case, it will cause the driver transformer to become relatively large since it must operate at such a low frequency. Figure 10-32 provides a method of achieving transformer isolation that can be switched at any frequency, while maintaining the isolation transformer very small.

A D flip-flop is used to provide alternating drives to the input of the primary drivers, similar to that of Fig. 10-31. The difference here is that a clock running at several megahertz is connected to the clock input to the flip-flop, so the transformer operating frequency is very high. That alone will limit the size of the transformer. The "set" and "reset" inputs of the flip-flop are tied together and driven by the control circuit at the actual switching frequency to be supplied to the MOSFET. When the control line is low, the transformer primary is driven at the high frequency. When it is high, both outputs go high, connecting both ends of the transformer primary to common, terminating drive to the MOSFET. On the secondary side of the transformer a full-wave bridge rectifies the high-frequency square wave at the secondary to a DC voltage. The capacitor, C_1, filters out the narrow negative edges that appear at the end of each high frequency half-cycle. So long as the high frequency pulses continue to appear at the bridge, DC voltage will be supplied to the gate of the MOSFET. As soon as the control signal goes high, primary drive is terminated, and C_1 begins to discharge through R_1. Once C_1 has discharged low enough to turn on Q_1, the gate of the MOSFET will be pulled down quickly, turning off the MOSFET. Capacitor C_1 must be charged and discharged each time the MOSFET is to be switched on or off, so it is important to keep the value of this part quite small. The value of R_1 is also important. If it is too large it will take a long time to discharge C_1 enough to turn on Q_1. If it is too small, it will load down the DC gate voltage.

As seen from these examples, it is relatively simple, compared to bipolar base drive, to turn power MOSFET devices on and off. The trick is to do it both fast and economically. If that is not achieved, then the whole advantage of using power MOSFETs over bipolar transistors is lost.

Power Converter Circuits

Several circuit configurations can be employed to achieve DC-to-AC power conversion. Each has its own set of advantages and disadvantages under certain conditions. This section will cover some basic approaches to various configurations along with a brief discussion of the operation of each. Figure 10-33 shows a full-bridge converter, output transformer, and output filter. The power switches are shown as switch elements for simplicity, but could be any type of device.

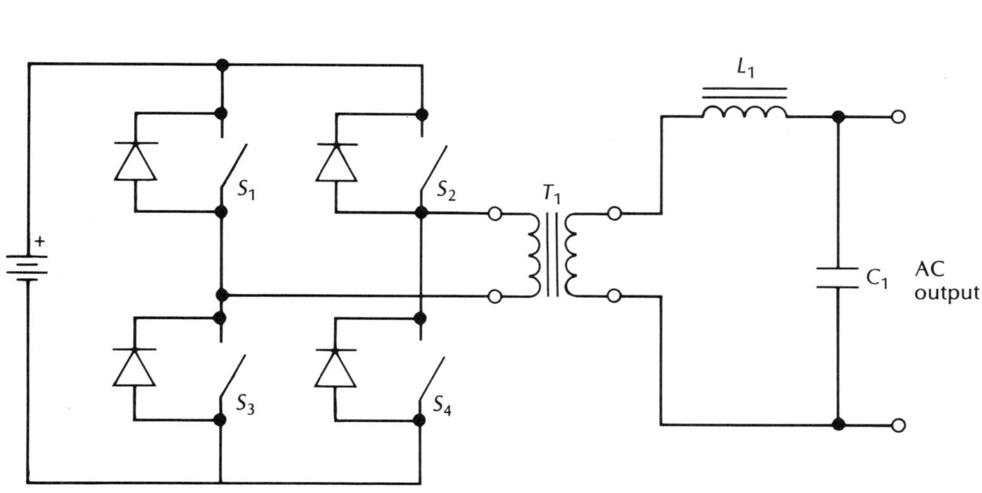

FIGURE 10-33 Full-bridge inverter.

Alternating-current voltage is impressed across the primary of T_1 by alternately switching between combinations of upper and lower devices. Switches S_1 and S_3 will conduct for a period, then S_2 and S_4 will conduct for the next. The voltage appearing at the transformer is then stepped up to the desired output voltage, filtered by L_1 and C_1, and delivered to the output. This arrangement has the advantage of operating with power devices that are voltage-rated very near the DC input voltage. When the output frequency is to be 60 Hz, for example, then the top devices can be switched, 180° out of phase with each other, at a 60-Hz rate, while the bottom devices are switched with PWM to get the desired wave shaping. Disadvantages to this circuit are a need to provide some form of level shifting, whether boot-strap or isolated drive, to operate the top devices. Each power device must be rated to handle the full current, and the fact that this current must travel through two devices in each direction doubles the voltage drop. This will have an effect of overall efficiency, particularly when low DC voltage is used.

The circuit shown in Fig. 10-34 is a half-bridge converter. It works in a similar manner as the full-bridge, but requires only half the power devices.

Alternately driving S_1 with PWM for half of the output period, then S_2 for the other half will provide the desired AC voltage at the primary of T_1. The advantage of this circuit over the full-bridge is improved efficiency and cost of power devices since only half as many devices are required to do the same job. The greatest disadvantage, particularly when a battery is used for the DC source, is the requirement for both a plus and minus supply. If batteries are used, then two 12-V batteries could be used for a total of 24 V. If the full-bridge circuit used a 24-V battery, then the half-bridge would require twice the current in the primary since it is switching to only 12 V for each half-

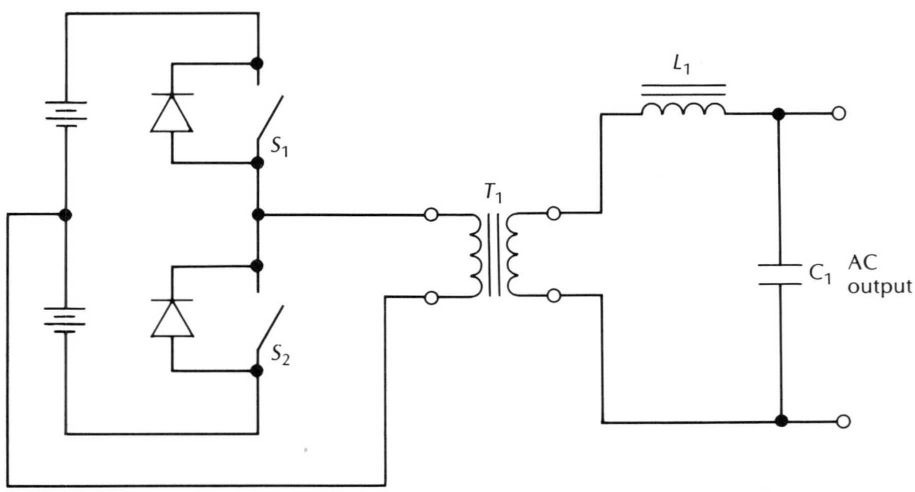

FIGURE 10-34 Half-bridge inverter.

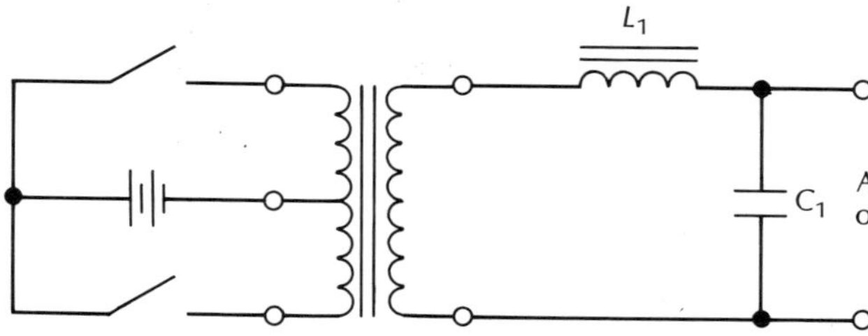

FIGURE 10-35 Push–pull inverter.

cycle. The same drive considerations must be made for half-bridge converters as was made for the full-bridge.

A push–pull converter as shown in Fig. 10-35 operates by alternately switching the power devices, just as in the half-bridge. The difference here is that T_1 is constructed with two full primaries in order to ensure the proper voltage transfer to the output. When S_1 is switched on, current flows from the DC source at the center tap of T_1 through half of the primary. When S_1 turns off and S_2 turns on, the current flows through the other half of the primary. The switching devices operate alternately, each being driven by PWM for half of the total output period. The advantage of this circuit over the others is that the drives can be referenced to control circuit common, with no need for level shifting. That alone will often make this a very desirable circuit topology. There are two main disadvantages. The first is the need for a larger power transformer, compared to the full-bridge circuit, to do the same job. Each half of the primary winding will have to handle the same current, given the same DC source, and will have the same voltage impressed on it as in the full-bridge. This circuit will require higher-voltage-rated power devices also. When S_2 is turned on, then the bottom end of the primary winding can be considered for practical purposes to be at 0 V. If a 24-V DC source is used, then the center of T_1 will be at 24 V. The autotransformer action of the primary will cause the top end of T_1 then to be at 48 V. So each power device will see double the DC source voltage during its off time. This circuit also requires very tight coupling of the two primary windings in order to minimize leakage inductance. As the leakage inductance between the primary windings increases, switching spikes will become a problem as the power device turns off. The sudden interruption of current flow in this inductance causes the voltage to rise sharply.

All the circuit topologies discussed above have one thing in common. The output transformer operates at power-line frequencies and can become quite large and expensive. The PWM drive to the power devices may be at high frequencies, but that really only helps to reduce the size of the output filter. There are only a few ways to prevent the need for the transformer. One is to use a boost converter to step a low DC voltage up to high voltage, and then apply a half- or full-bridge converter with no output transformer. This works quite well, but is really two independent power converters operating, each

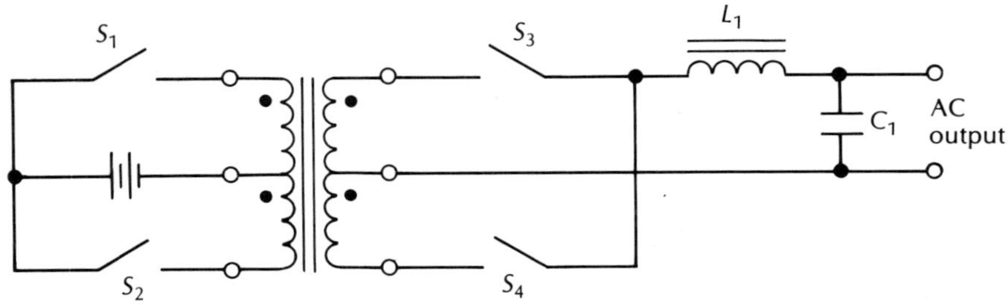

FIGURE 10-36 Phase-controlled inverter utilizing high-frequency transformer.

with their own PWM control and feedback loops. It increases the chances of failure to have more converters operating at the same time and reduces the overall conversion efficiency. The circuit shown in Fig. 10-36 provides a means of ensuring high-frequency switching and stepping up the voltage from a low-voltage DC source, without the need for a low-frequency transformer, all in a single converter. Transformer T_1 is arranged similar to that in the push–pull converter described above. In this case, though, the primary is switched alternately by S_1 and S_2 at a high frequency. So the first advantage is immediately realized, T_1 can be made quite small because of the high frequency it is operated at. The primary switches are not driven with pulse width modulation in this circuit. They are simply switched with a 50–50 duty cycle at the operating frequency. The secondary of T_1 is designed to provide the desired peak output voltage from each half of the winding. Switches S_3 and S_4 also are to be switched at a 50–50 duty cycle, at the operating frequency. But their operation is to be phase-shifted with respect to the drive to the primary switches. By following the flow of current through the primary winding, and following the polarity of the voltage transformed of the secondary, it can be seen that when S_1 is conducting, and S_3 is also conducting, but phase-shifted 90°, there will be zero output. Then as the drive to S_3 is advanced or retarded from 90°, the output voltage will go plus or minus. By modulating the drive to the switches on the secondary, with sine-weighting, and filtering the pulses, a very nice sine-wave voltage can be realized at the output. All this is accomplished with a single control circuit that does not have to provide PWM, simple drives to all four switching devices, and a single high-frequency transformer. Care must be taken in the construction of the transformer to minimize leakage inductances, otherwise very high voltage spikes will appear across the secondary switches. The secondary switches normally would see the full peak to peak winding voltage, anyway, so they must be high voltage devices. But the additional spiking could lead to serious problems.

CHAPTER 11
Thermal Management

Electronic power conversion equipment relies on the flow and control of electrical current to perform its task. Whenever electrical current flows through a resistive element, heat is generated in that element. An increase in the current or resistance produces an increase in the amount of heat that is generated in the element. The heat continues to be generated as long as the current continues to flow. As the heat builds up, the temperature of the resistive element starts to rise, unless the heat can find a path that carries it away from the element. If the heat flow path is poor, the temperature may continue to rise until the resistive element is destroyed and the current stops flowing. If the heat flow path is good, the temperature may rise until it stabilizes at a point where the heat flowing away from the element is equal to the heat generated by the electrical current flowing in the element.

Heat is generated by the flow of electrical current in electronic component parts such as resistors, diodes, transistors, and transformers—most of the parts found in a typical power converter design. The temperatures must be controlled on every component to ensure a reliable product. If the operating temperatures become too high, failures may occur. Thermal problems may show up as an out of tolerance condition or as a catastrophic failure in which components smoke, burn, and sometimes physically blow apart. Heat always flows from the hot area to the cool area. Since the electronic components are usually the source of the heat, the electronic components will usually be the hottest spots in an electronic system. The basic problem of thermal management in any electronic equipment is, therefore, the removal of internally generated heat by providing a good heat flow path from the heat sources to an ultimate sink, which is often the surrounding ambient air. There are three basic methods by which heat can be transferred: conduction, convection, and radiation. It is important to have an understanding, at least in principle, of these three heat-transfer mechanisms in order to properly design power conversion products, and allow the components to achieve an acceptable life span in operation. Consider that a 200-W power supply operating at 80% efficiency will require 250 W of input power. The difference between the input and output power is caused by the losses within the power converter and is reflected in heat generation. As an experiment, take a 40- or 60-W lightbulb and place it inside a small enclosure with no ventilation and monitor the temperature within. In a short time the temperature can reach levels that will degrade or damage electronic components, reducing their effective life. Now by simply adding a few vent holes in the bottom and top of the enclosure, and placing the enclosure in a position that will allow convection air currents to rise through the enclosure, the temperature will drop significantly. Without proper thermal management the mean time between failure (MTBF) of any power conversion product will be reduced.

The following discussions will detail the transfer of heat through conduction and convection, and present examples of how to apply the principles. Thermal radiation is the transfer of energy by electromagnetic waves that are produced by bodies because of their temperature. A hot body radiates energy in all directions. When this energy strikes another body, the part that is absorbed is transformed into heat. Most electronic systems make use of all three basic methods of heat transfer to some extent, even though one method may dominate the design. For example, an electronic enclosure cooled by forced convection might utilize a fan to draw air over components mounted on printed-circuit boards. The greatest amount of heat is picked up by forced convection as the cooling air passes over the individual components that are mounted on the boards. However, some of the heat is conducted to the backside of the printed circuit boards, through the components electrical lead wires. Since the cooling air passes over both surfaces of the printed circuit boards, the conduction of heat to the backside of the boards provides additional surface area for improved cooling. In addition, some of the heat is radiated from the hot components to the surrounding chassis walls and to the cooler spots on adjacent printed circuit boards. This helps reduce the component hot-spot temperatures. Attempting to rely too much on the process of radiation transfer to cool components is risky at best, and generally in commercial power conversion equipment it is best to design around conduction and convection and presume that whatever radiation help you can get is a plus. For this reason there will be little more said on the topic of radiation cooling.

Cooling by Conduction

Conduction is the transfer of kinetic energy from one molecule to another. In an opaque solid it is the only method of heat transfer, where heat flows from the hot areas to the cooler areas of the solid. Heat conduction also occurs in gases and liquids, but the amount of heat transferred is usually smaller for the same geometry. Conduction is the method by which heat is transferred from the semiconductor chip within a power transistor the case within which it is bonded. All solid-state devices, such as power transistors, MOSFETs, and diodes, contain a silicon or germanium chip. As current passes through the chip

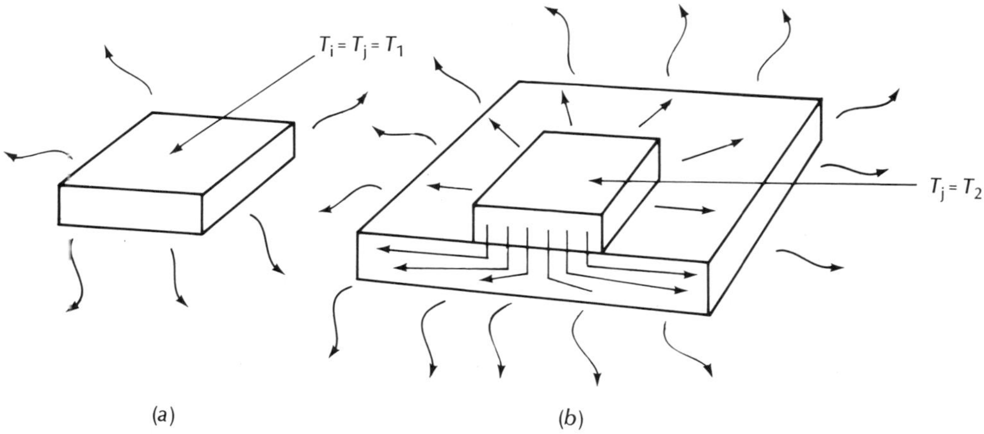

FIGURE 11-1 (*a*) Internally generated heat radiating from an isolated block (T_1 = surface temperature of the block); (*b*) heat flows from the block into an attached heat sink; the entire assembly then radiates heat, leaving $T_2 < T_1$.

it dissipates power due to I^2R loss. The current density and power dissipation are greatest in regions called *junctions*, where the current control takes place. Controlling the junction temperature is critical to a reliable design.

Figure 11-1 depicts a model of a transistor chip as the power generating block and $T_i = T_j$ as the semiconductor chip temperature. In Fig. 11-1*b* the chip is bonded to a heat-sink mounting surface, part of the case of the transistor. Presume that this transistor chip is active and conducting current, while dissipating 1 W of power. This power dissipation will cause the chip temperature to increase until the amount of heat leaving the chip is the same as the amount of heat generated as a result of the power dissipation. The two quantities are totally independent of each other. The level of power dissipation is a function of the power applied and the resistance-to-carrier flow in the device. The heat leaving is a function of the size and configuration of the chip and the temperature rise above the surroundings. Equilibrium between the power in the heat out will be attained only when the temperature has risen to a point exactly high enough to allow dissipation of 1 W. The smaller the chip, the higher the temperature required to dissipate the 1 W. Chip temperature could easily reach several hundred degrees Celsius, depending on the size of the chip.

Reliable operation of a silicon chip will not allow temperatures to exceed 200°C, and even operating at that temperature is quite risky. While it has been shown that operation to even 300°C is possible, the reliability suffers dramatically once 200°C is exceeded. Most designers will try to maintain a worst-case junction temperature that will not exceed 50–60% of the manufacturer's specified maximum operating temperature. One reason, in addition to the sheer reliability consideration, is the effect that temperature can have on so many of the parameters of the part. For example, a particular power MOSFET device may be rated for continuous current operation of 9 A at 25°C case temperature. This same device may then be rated only to 6 A at 100°C case temperature, while the maximum junction temperature is rated to 150°C. If this part is to be operated in a room environment, it is likely the case temperature will exceed 25°C, unless great expense is incurred utilizing exotic cooling methods. Either natural or forced-convection cooling will never allow the part to be any cooler than its surrounding environment, in this case the room temperature. So in this case the 9-A rating will be derated by whatever case temperature at which the final design allows the device to operate.

Getting back to the 1-W dissipation in the chip above, if it is necessary to hold the junction temperature down to a level below where the device itself would operate at, then some external means of taking control of the temperature must be implemented. By mounting the device to a heat sink the chip can now conduct heat out through its package to the external heat sink and the dependence on the chip geometry as the sole means of dissipating the heat is removed. The chip is no longer constrained to radiate away its energy as a function only of its own surface area and temperature. Because of the increased dissipation area, the heat-sink assembly will transfer the same amount of heat at a lower temperature and the device will operate within its allowable temperature limits.

In solid-state devices the semiconductor chips are bonded onto metals and ceramics that act as heat sinks. The device leads also aid in conducting heat away from the chip. The whole assembly is usually packaged in a case that is on the order of 1000 times the size of the actual chip. The whole device is transferring heat away into the surroundings. This is the device manufacturer's package, commonly called the "case." When case temperature is specified, usually as T_C, it is the temperature of this enclosure around the chip, and not the actual chip temperature. The manufacturer can determine the temperature of the chip, or more specifically the junction temperature, at any given case temperature and then specifies a parameter θ_{jc} that describes the thermal impedance between the junction of the chip and the case of the device. Knowing this, the manufacturer can then predict the junction temperature based on measured case temperature.

Thermal resistance is the parameter most often used to characterize heat sink performance. The term is the reciprocal of the heat-transfer term "thermal conductance." "Thermal resistance" is the preferred choice in heat sink selection where an electrical model is employed for the solution of simple problems. Heat flux q is expressed most commonly in watts when dealing with electrical power dissipation. There are various causes why the flow of heat may be restricted or impeded. This restriction or impedance is the thermal resistance. When thermal resistance is encountered, a driving potential is required to surmount it. This driving potential is the temperature rise ΔT. A relationship between the thermal resistance θ and the temperature difference ΔT exists for a constant heat flow q.

$$\Delta T = q\theta \tag{11.1}$$

where $\Delta T = T_1 - T_2$ (°C), q = watts, and θ = °C/W. Note that the heat being transferred across a thermal resistance requires a temperature difference equal to the product of $q\theta$ to maintain the flow of heat.

For purposes of an electrical model, consider Ohm's Law:

$$\Delta V = IR \tag{11.2}$$

The basic law says that current I passing through a resistance R gives rise to a potential difference ΔV equal to the product IR. Such a model is depicted in Fig. 11-2. The thermal resistance is a physical property. Thermal resistance θ_{jc} represents the summation of the internal resistance of the chip, the resistance between the chip and the case, and the internal resistance of the case. These properties are built into the device and are not a design variable. Figure 11-3 depicts how a power diode might appear mechanically, and schematically. Resistance θ_{cs} is a complex summation of surface and interstitial (gap) resistance known as the mounting or interface thermal resistance. This gap at the interface is caused by the uneven surfaces being joined together. The resultant air gap causes thermal resistance that can be reduced through the use of special compounds designed for just this purpose. Resistance θ_{cs} may also include the overall resistance of insulators placed between case and heat sink; θ_{cs} may be minimized by following strict assembly procedures. Resistance θ_{ca} is very high relative to the total resistance and, because it is a parallel resistance path, is usually ignored. The variable for the designer is the thermal resistance from heat sink to ambient, θ_{sa}. Controlling θ_{sa} with q fixed determines the sink

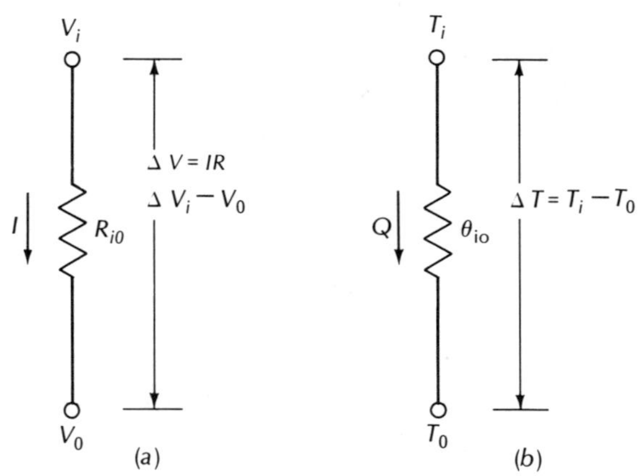

FIGURE 11-2 Schematic representation of similiarity between voltage and thermal drop across impedance.

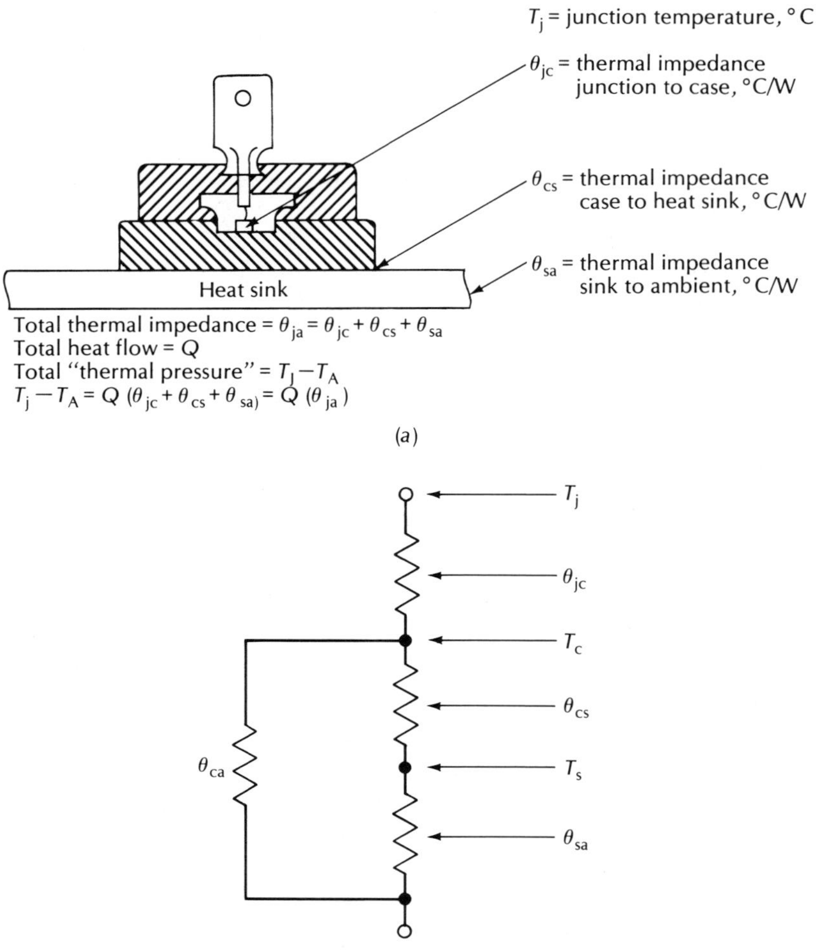

FIGURE 11-3 Mechanical and electrical models depicting thermal path.

temperature rise above ambient in the heat path. It can shift all the other temperatures up or down. Changing θ_{sa} also varies the overall thermal resistance θ_{ja}, which, in turn, determines to what extent the input power q_i could be increased while not exceeding a specified junction temperature T_j.

Design Example

It is determined that one power transistor is dissipating 60 W. Its junction temperature T_j is not to exceed 180°C, at an ambient temperature of 40°C maximum. The thermal resistance listed by the manufacturer for the junction to case θ_{jc} is 0.6°C/W. The thermal resistance at the mounting interface is

0.2°C/W. Calculate the required thermal resistance of the heat sink θ_{sa} to maintain the specified junction temperature. The first step is to determine the maximum allowable junction temperature change ΔT:

$$\Delta T = 180°\text{C} - 40°\text{C} = 140°\text{C maximum} \tag{11.3}$$

Next calculate the total thermal resistance in the assembly:

$$\begin{aligned} \theta_{total} &= \theta_{jc} + \theta_{cs} + \theta_{sa} \\ &= 0.6 + 0.2 + \theta_{sa} \\ &= 0.8\ °\text{C/W} \end{aligned} \tag{11.4}$$

From (11.1):

$$\begin{aligned} \Delta T = q\theta &= 60\ (0.8 + \theta_{sa}) \\ &= 48°\text{C} + 60\theta_{sa} = 140°\text{C} \end{aligned}$$

Solving for θ_{sa}:

$$\theta_{sa} = 1.53°\text{C/W}$$

Heat sink manufacturers generally display their extruded heat sinks in cross section, illustrating the shape and dimensions, and provide vital data for a 3-in section. On observation it can be seen that the lower the thermal resistance of the heat sink, the better it will work for the application described above. Selecting a heat sink that provides a thermal resistance of 1.53°C/W or less will yield the desired thermal results. As an exercise it can also be found that if the single transistor of this example were divided up into several transistors, still dissipating a total of 60 W, a heat sink of higher thermal resistance could be selected. Generally a higher thermal resistance implies a smaller size. Solve for the thermal paths as parallel paths, just as you would solve for total resistance in parallel resistors.

While the heat sink provides the thermal resistance solved for above, it should be noted that this is specified in free air, and with no air flow, beyond the natural convection process. How the actual heat sink selected will perform will have much to do with how it is applied. Natural or forced convection will play a role in determining the effectiveness of the heat sink, and, as will be demonstrated in the next section, a little bit of air movement can go a long way toward reducing the thermal resistance of a given heat sink, thus allowing the designer to select a smaller heat sink for the same job. One other thing to consider before leaving the topic of heat conduction is that the conduction process does not necessarily stop with the addition of a heat sink to the semiconductor case. If the heat sink itself is properly secured to a good heat conductor, such as an aluminum or steel enclosure, then two parallel processes are at work cooling the semiconductor. The conduction of heat into the heat sink is followed by conduction from the heat sink into the enclosure. Added to this then is the convection process cooling both the heat sink and the enclosure, provided of course that the enclosure is mounted in free air. It can be seen that there are many paths for cooling components, and by thinking through the problem there are many solutions.

Cooling by Convection

Convection is the transfer of heat by the mixing action of fluids. When the mixing is due entirely to temperature differences within the fluid, resulting in different densities, the action is known as *natural convection*. When the mixing is produced by mechanical means, such as a fan, the action is known as *forced convection*.

Heat transfer by natural convection will occur when there is a change in the density of the fluid, which can be a gas or a liquid. Fluids tend to expand as they are heated, which results in a reduced density. In a gravity field, the fluid, which has a lower density, is lighter and therefore rises, creating a movement in the fluid that is called *convection*. This movement or convection permits the fluid to pick up heat and carry it away. When the fluid movement is produced by only a difference in the fluid density, the process is called *natural* or *free convection*. When the fluid movement is mechanically induced by a fan, for instance, the process is called *forced convection* or *forced circulation*. The driving force for natural convection is not very great because it depends on the density change in the fluid. Therefore, any small obstacle or resistance in the flow path will sharply reduce the fluid flow rate and, therefore, the cooling rate. An example of this is the effect of lighting a fire in a fireplace. With the flue open, and a sufficient amount of leakage for outside air to enter the room, the smoke from the fire will go up the chimney. This is due to natural convection because the fire is heating the air in the fireplace, causing the air to rise, and drawing in cool air from the room to replace it. Now, if the room were sealed off so no more outside air could enter, the smoke from the fire would soon fill the room because the convection process would have stopped.

Heat transfer by natural convection is more efficient if a well-defined, clear, unobstructed flow path is made available. Otherwise, the amount of heat carried away may be much less than anticipated and the resulting temperatures may be much higher than expected. Natural-convection heat transfer is extremely easy to use because no additional equipment is required. It is also very reliable since there are no moving parts, such as a fan, to wear out.

On the other hand, as will be discussed in the following paragraphs, adding a fan as part of the cooling system can have several advantages. Aside from the fact that smaller heat sinks will now work in place of larger ones, there is less restriction on the shape and location of the heat sinks. It is also often valuable to be able to remove heat from the enclosure to cool components that are not easily heat sunk, such as transformers and other passive components that, because of their power dissipation, may have to be larger if no fan is used. But care should be exercised in placing the fan, and shaping the path that the cooling air must travel, in order to obtain maximum utilization. As cooling air flows through the enclosure, it will experience friction and dynamic losses. There are many ways of expressing these losses. One convenient method is to express the static pressure losses in terms of the velocity head for the air at some specific point. These losses can be expressed as some ratio of the velocity head at the fan inlet. The loss at a 90° bend can be expressed as some ratio of the velocity head at that bend, which is related to the velocity of the air flowing through the bend. A loss in sudden expansion is expressed as a ratio of the velocity head just before the expansion. A typical sudden expansion will result in a loss of one velocity head, because the air velocity at the expansion is sharply reduced. Most of the energy required to provide that velocity at the point just before the expansion is lost. That energy was originally supplied by the fan. If more energy is lost, a larger fan is required to supply that energy. If the losses in an air-flow system can be reduced, by avoiding such things as sharp turns and sudden expansions, a smaller fan can be used. The whole subject of forced air flow and how the design of an enclosure impacts the efficiency of a cooling system is beyond the scope of this book. Generally, unless an experienced mechanical engineer with this sort of background is utilized in the design of a forced air cooling system it is either "over-killed" from the outset or suffers from multiple redesigns to achieve the required performance.

Figure 11-4 illustrates the effect of forced air flow over a heat sink. On this graph are two sets of data for a particular heat sink. One curve, traveling from the lower left to the upper right, is the natural-convection curve of power dissipated versus temperature rise. The left scale of the graph gives the temperature difference between the hottest point on the heat sink T_s and the surrounding air T_a, or $\Delta T_{sa} = T_s - T_a$, in degrees Celsius. The bottom scale of the graph gives the power dissipated by the device. The curve from the upper left to the lower right is the forced-convection curve of air velocity versus thermal resistance. Air velocity is shown on the top scale and the thermal resistance with forced convection is shown on the right-hand scale.

Each thermal resistance curve is unique. Each heat sink has its own curve for each mode of heat transfer. The test condition of the heat sink must be defined. Figure 11-4 describes the heat sink and conditions of measurement. For any particular power dissipated, a vertical line may be drawn from that point intersecting the natural-convection curve. A horizontal line to the left from the point of intersection gives the ΔT_{sa}, or sink temperature rise above ambient. This is the temperature rise above ambient that the heat sink attains directly under the device for that particular power dissipated. Note that the slope of the curve is not generally a straight line. The slope, $\Delta T_{sa}/W(\theta_{sa})$,

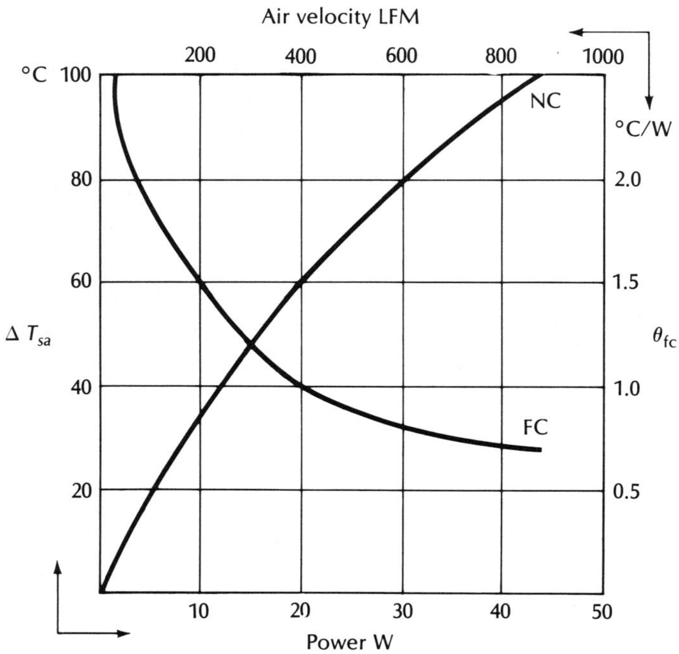

FIGURE 11-4 Typical heat-sink characteristic curve.

will start out high, flatten out in the region of $\Delta T = 50°C$, and drop off above $\Delta T = 60–80°C$. Since the curve is straightest in the region of 50°C, the thermal resistance is usually specified at $\Delta T = 50°C$. However, advances in device construction permit temperature values in the 100°C range, with sink temperatures to 150°C. In these instances, the curve should be utilized in this region rather than using the specified θ_{sa} at $\Delta T = 50°C$. Examples of θ_{sa} for natural conection are shown in Fig. 11-4. Also shown in Fig. 11-4 is the forced-convection curve. This curve is independent of the power applied provided the power is greater than that needed for $\Delta T = 50°C$ in natural convection. In other words, the forced-convection curve would be the same in Fig. 11-4 if the test were run at 20, 30, or 40 W. The forced-convection curve of Fig. 11-4 was run at $q = 40$ W. Air was blown over the heat sink and ΔT_{sa} measured. Then ΔT_{sa} was divided by $P = 40$ W to give θ_{fc} for that air flow. At 500 lineal feet per minute (LFM), for example, $\theta_{fc} = 1°C/W$. Since $P = 40$ W, $\Delta T = 40°C$. If $P = 30$ W instead, ΔT would be 30°C, and $\Delta T/P = 30/30$ would still equal 1°C/W.

Design Example

1. Given a heat sink with the characteristic curve of that shown in Fig. 11-4, find the thermal resistance of the heat sink with 200 LFM air flow. Begin by following the vertical line from 200 LFM down until it intersects with

the forced-convection curve. Then draw a horizontal line from that intersection to the right-hand scale and mark the result at $\approx$ 1.5°C/W.

2. What air velocity is required to achieve 1°C/W? Draw a horizontal line from 1°C/W on the right-hand scale, left until it intersects with the forced-convection curve. Now draw a vertical line from this intersection up to find the required air flow to be 400 LFM. Often the air flow is specified for fans and heat sinks in cubic feet per minute (CFM). If the heat-sink characteristic curve and fan both specify the same air-flow units, then the curve is used as described above. When a conversion between LFM and CFM is required, then use the following:

$$\text{LFM} = \frac{\text{CFM}}{\text{cross section of flow (ft}^2\text{)}} \tag{11.5}$$

These characteristic curves give the thermal resistance of a heat sink generally with only one semiconductor mounted in the center. In the typical heat-sink application, two or more devices may be mounted on one heat sink. This spreads the load and the true thermal resistance is only 70% of that shown in the curve for natural convection. For heat sinks with constant cross section, a distributed power load may lead to some simplifications. When the power is evenly distributed over the heat sink:

a. Divide the total area by the number of devices.
b. Determine the thermal resistance for that cross-sectional area.
c. Multiply by the 70% factor.
d. Divide by the number of devices to obtain the effective overall thermal resistance.

This is a great generalization and will not be valid for nonsymmetrical heat sinks or those with input surfaces of varying thickness.

3. If a fan is used to duct cooling air at 100 CFM into an opening 2 in high and 12 in wide, what is the air velocity? First solve for the area of the opening:

$$\text{Area} = 2 \times 12 = 24 \text{ in}^2 = \tfrac{1}{6} \text{ ft}^2$$

Using (11.5), solve for LFM:

$$\text{LFM} = 100 \text{ CFM} \div \tfrac{1}{6} = 600 \text{ feet per minute}$$

Each heat sink found in a manufacturer's data book will usually have its own unique characteristic curve depicting the performance in thermal impedance for given power and air flow. Experience has shown these curves to be quite accurate and very useful in predicting the operating temperature of power transistors. Some difficulty is encountered when using the forced air curve, however, in that it is often difficult to know just what the actual air flow in a system will be prior to actually building the first prototype enclosure. Back pressure and other considerations can greatly reduce the volume of air that a particular fan can push through an enclosure, thus having a reduced effect on cooling. It pays to use caution and be a little conservative when planning to use forced convection to cool power components. Where it is practical to do so, given size, weight, and other considerations, natural convection is the better choice.

CHAPTER 12

Filter Design

The purpose of adding a filter to the output of any DC-to-AC power converter (inverter) is to improve the waveshape. In its simplest form an inverter will switch square-wave voltage at the desired output frequency, resulting in a square wave with a peak voltage required to operate the load. This square wave will contain only odd order harmonics (3rd, 5th, 7th, 9th, . . . , *n*th), and the total harmonic content will be about 47%. If the load doesn't care what the waveform looks like, and is not bothered by harmonic content, there is no need to add the cost and weight of a lowpass filter to the output of the inverter. Unfortunately the load is not always defined, and we must consider that there will be motors or other elements of the load that will be disturbed by the distortion. Motors will run hotter with higher harmonic content, and some instrumentation may give inaccurate readings.

Before considering the output filter, it is important to give some consideration to the waveform generated ahead of the filter. There are methods that will help minimize the harmonic content of the waveform before it gets to the filter, thus reducing the size and cost of the filter. Chapter 13 discusses in more detail the area of control circuitry and waveform generation.

Typically the output filter used is a lowpass configuration made up of one or more inductors and one or more capacitors. A lowpass filter is designed to be frequency-selective, in that it will pass low frequencies and block high frequencies. Figure 12-1 depicts an ideal amplitude response of a lowpass filter. The dotted line in Fig. 12-1 shows the ideal response, while the more rounded corners of the sloped curve form the realizable curve. A lowpass filter has a single passband, $0 < \omega < \omega_c$, and a single stopband, $\omega > \omega_c$. The frequency ω_c, expressed in hertz as $f_c = \omega_c/2\pi$, separates the two bands and is called the cutoff frequency. The ideal response of Fig. 12-1 is impossible to realize since the transfer function of an actual filter is a ratio of polynomials and its amplitude cannot possess the sharp corners of the ideal response. In the realizable case, represented by the solid line in Fig. 12-1, the division between the passband and the stopband is not clearly distinguishable, and the cutoff frequency is defined as the point at which $H(j\omega)$ is $1/\sqrt{2} = 0.707$ times its maximum value, shown as A in Fig. 12-1. In decibels, the cutoff frequency is the point at which the amplitude is 3 dB below its maximum

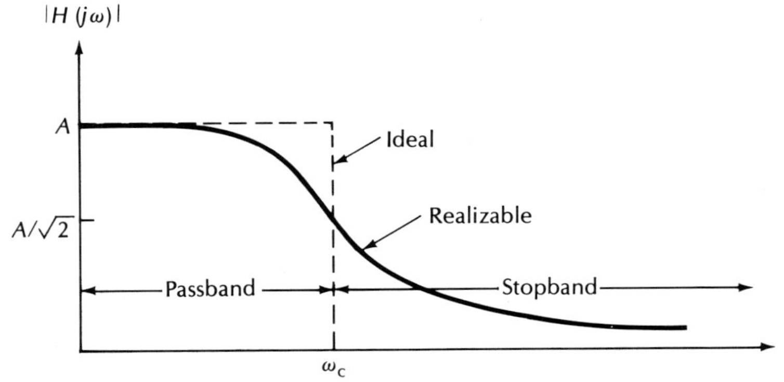

FIGURE 12-1 Ideal and realizable lowpass responses.

value. The gain of a lowpass filter is the value of its amplitude at zero frequency, or equivalently, gain $= H(0)$, where $H(s)$ is the transfer function. The simplest transfer function that approximates the ideal lowpass characteristic is the ratio of a constant to a polynomial. It is called an all-pole function because all its poles, and none of its zeros, are finite. More on this will follow.

Image Impedance

By combining building blocks, referred to as *half-sections* and *full-sections*, as shown in Fig. 12-2, image-parameter filters can be constructed. Element value equations are based on the assumption that the network is terminated in impedances that satisfy the following expressions:

$$Z_T = \sqrt{Z_1 Z_2 + \frac{Z_1^2}{4}} \tag{12.1}$$

$$Z_\pi = \frac{Z_1 Z_2}{\sqrt{Z_1 Z_2 + \frac{Z_1^2}{4}}} \tag{12.2}$$

where Z_T and Z_π are referred to as *image impedances*. The input impedance of the network is also defined by these expressions, when the network's output is terminated with Z_T or Z_π, as illustrated in Fig. 12-2. Impedances Z_T and Z_π may change rather abruptly with frequency since Z_1 and Z_2 are reactive elements. This makes precise terminating of image-parameter filter almost impossible. This is a basic limitation to the image-parameter concept. At some frequency within the filter passband the image impedance becomes purely resistive. Terminating the filter in this nominal value R approximates the image impedance over much of the passband. Still the inability to terminate

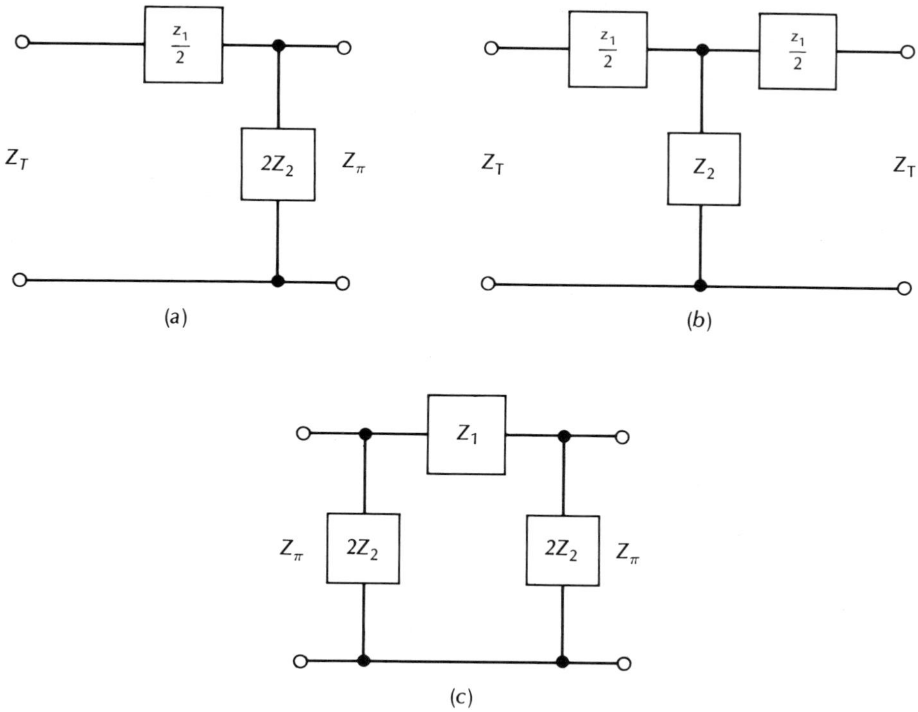

FIGURE 12-2 Filter building blocks: (*a*) half section; (*b*) full T section; (*c*) full π section.

image-parameter filters in their exact image impedance can result in large deviations of filter response from the theoretical predictions.

Constant-*k* Lowpass Filter

A constant-*k* lowpass half-section and its frequency response are shown in Fig. 12-3. The element values are computed by

$$L_k = \frac{R}{2\pi f_c} \text{ henrys} \tag{12.3}$$

$$C_k = \frac{1}{2\pi f_c R} \text{ farads} \tag{12.4}$$

where f_c, as defined earlier, is the cutoff frequency. The ultimate rate of roll-off of constant-*k* low-pass filters is limited to 6 dB per octave for each reactive element. A filter made up of two coils and three capacitors eventually rolls off at 30 dB per octave.

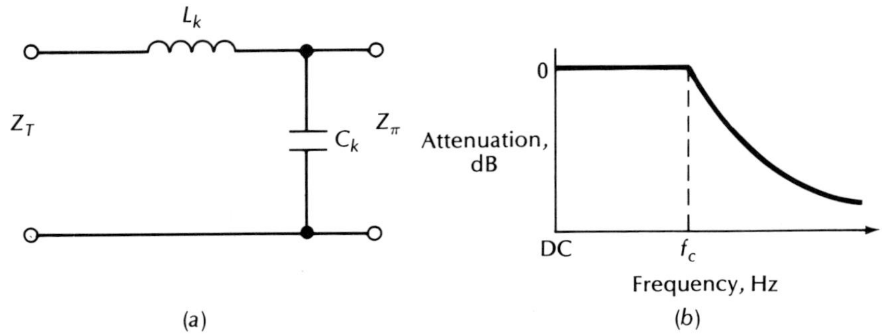

FIGURE 12-3 Constant-*k* lowpass filter half section.

Design Example

Multiple constant-*k* lowpass half-sections are to be combined to obtain an ultimate roll-off of 24 dB per octave. The desired cutoff frequency, f_c, is 1 kHz, and the load resistance R is 600 Ω.

Using (12.3) and (12.4), compute L_k and C_k:

$$L_k = \frac{R}{2\pi f_c} = .0955 \text{ H} \tag{12.5}$$

$$C_k = \frac{1}{2\pi f_c R} = 0.265\ \mu\text{F} \tag{12.6}$$

Remembering that each reactive element will provide 6 dB of roll-off, combine four half-sections, as illustrated in Fig. 12-4*a*, to achieve the desired 24 dB roll-off. Combining the values of the filter in Fig. 12-4*a* will result in a filter as depicted in Fig. 12-4*b*. This final circuit will have a passband behavior that may not be suitable for many applications as a result of the inherent deficiencies of the image-parameter concept. It is seen from this example, however, that it is relatively easy to compute and combine element values to reach the design goal, which is one of the few attributes of the image-parameter method.

m-Derived Lowpass Filter

Compared to the constant-*k* filter, the *m*-derived lowpass filter can provide a steeper rate of roll-off, with fewer circuit elements. An *m*-derived lowpass filter section has a transmission zero. This is a notch or resonance in the stopband having near infinite rejection. This notch can be located wherever a very steep rate of roll-off is required. Figure 12-5 shows an *m*-derived lowpass filter

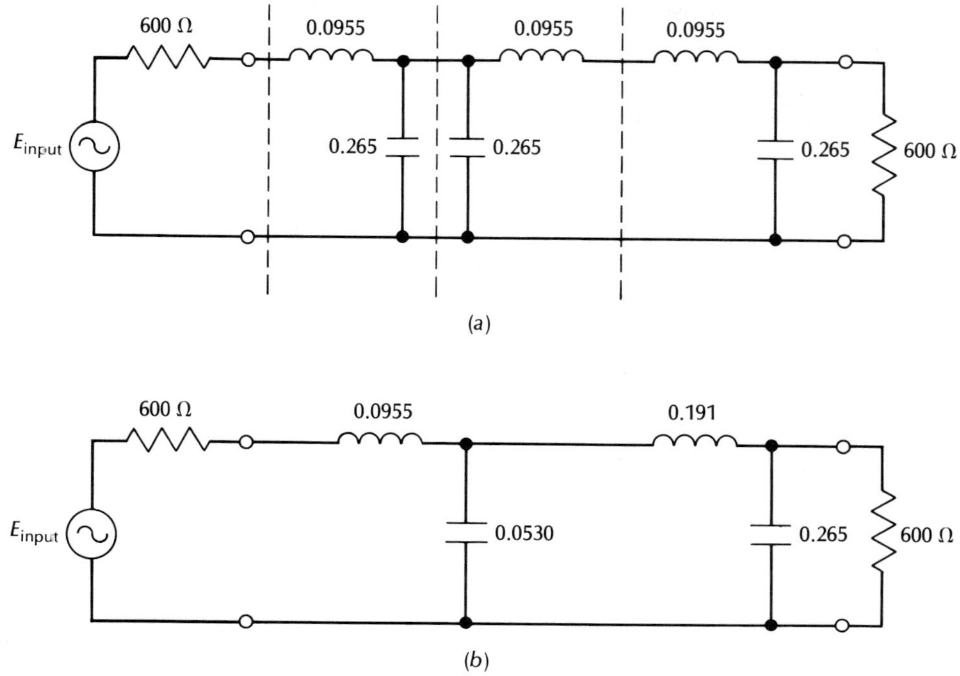

FIGURE 12-4 Example filter: (*a*) lowpass filter before combining elements; (*b*) final design.

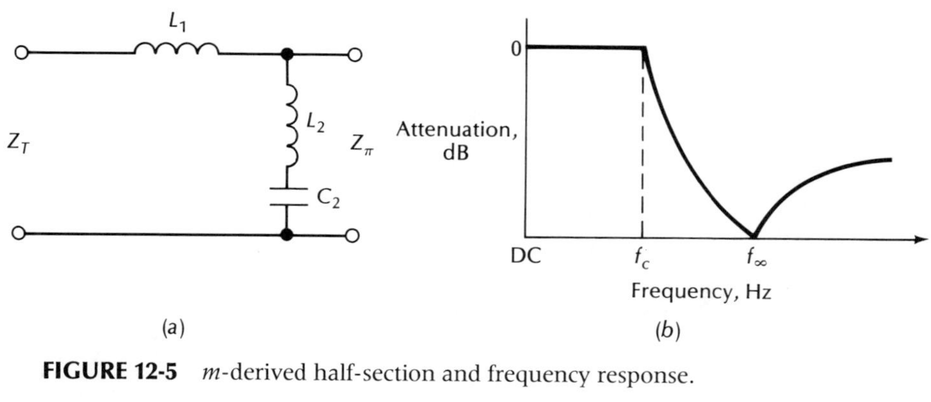

FIGURE 12-5 *m*-derived half-section and frequency response.

half-section and its frequency response. Although the notch, or transmission zero, provides a steeper rate of descent from the passband into the stopband, a return lobe, also known as a *comeback,* occurs beyond the notch. The amplitude of this lobe increases as the notch is placed closer to the cutoff frequency. It is possible to reduce the amplitude of the come-back lobe by combining constant-*k* and *m*-derived sections.

The following equations are used for m-derived half section elements:

$$m = \sqrt{1 - \frac{f_c^2}{f_\infty^2}} \tag{12.7}$$

$$L_1 = mL_k \tag{12.8}$$

$$L_2 = \frac{1 - m^2}{m} L_k \tag{12.9}$$

$$C_2 = mC_k \tag{12.10}$$

The effect of the design parameter m on the image impedance is plotted in Fig. 12-6. The flattest impedance over the greatest part of the passband occurs when $m = 0.6$. This value results in a good match to a resistive load; so image-parameter filters are usually terminated with $m = 0.6$ half-sections. It is important to recognize that m-derived sections provide two equally important functions. They provide a suitable match between the image impedance of constant-k sections and a resistive load termination, and also increase the rate of descent into the stopband. This added benefit of producing an infinite rejection notch at a desired frequency is helpful when trying to reduce the high-frequency content, caused by PWM, riding on a 60-Hz sine wave. The following example demonstrates the design of an image-parameter lowpass filter using both constant-k and m-derived sections.

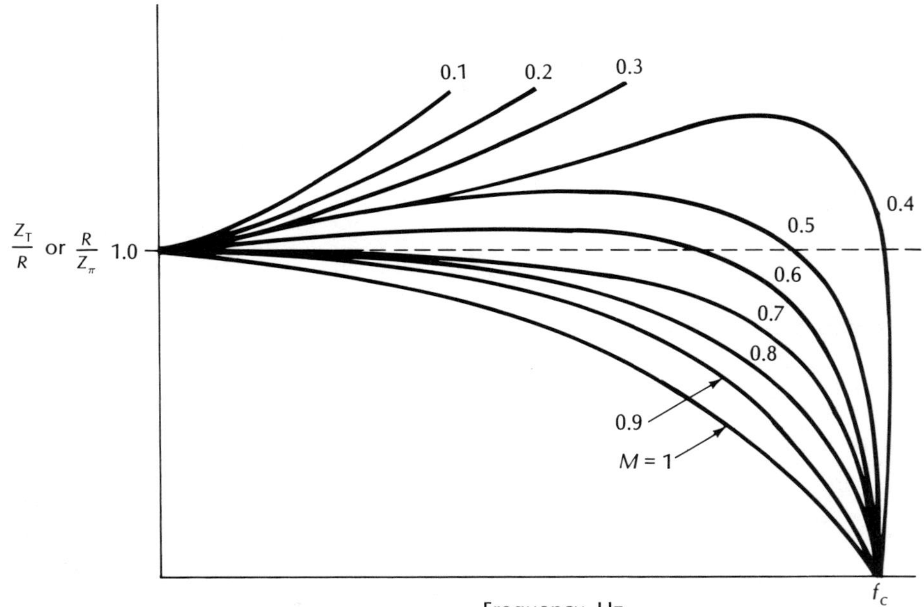

FIGURE 12-6 Image impedance versus m.

Design Example

A low-pass filter is required to have a cutoff frequency f_c of 5 kHz and a notch frequency f_∞ of 10 kHz. The filter will be terminated with a 600-Ω resistor R. The desired result can be achieved by using a constant-k full-T midsection, an m-derived full-T midsection, and m-derived end sections for terminating.

Begin by computing the constant-k midsection:

$$L_k = \frac{R}{2\pi f_c} = 0.0191 \text{ H} \tag{12.11}$$

$$C_k = \frac{1}{2\pi f_c R} = 0.0531\ \mu\text{F} \tag{12.12}$$

Next compute the m-derived midsection:

$$m = \sqrt{1 - \frac{f_c^2}{f_\infty^2}} = 0.866 \tag{12.13}$$

$$L_1 = mL_k = 0.0165 \text{ H} \tag{12.14}$$

$$L_2 = \frac{1 - m^2}{m} L_k = 0.00552 \text{ H} \tag{12.15}$$

$$C_2 = mC_k = 0.046\ \mu\text{F} \tag{12.16}$$

Finally, compute the m-derived end sections using an $m = 0.6$ to obtain the flattest response:

$$L_1 = mL_k = 0.0115 \text{ H} \tag{12.17}$$

$$L_2 = \frac{1 - m^2}{m} L_k = 0.0204 \text{ H} \tag{12.18}$$

$$C_2 = mC_k = 0.0319\ \mu\text{F} \tag{12.19}$$

The resulting circuit is show in Fig. 12-7*a*. By combining elements, the filter can be reduced to that shown in Fig. 12-7*b*.

Pole-Zero Concept

A generalized filter is shown in Fig. 12-8. The filter block may consist of inductors, capacitors, resistors, and possibly active elements such as operational amplifiers and transistors. The terminations shown are a voltage source E_s, a source resistance R_s, and a load resistor R_L. The circuit equations for the network of Fig. 12-8 can be written by using circuit-analysis techniques.

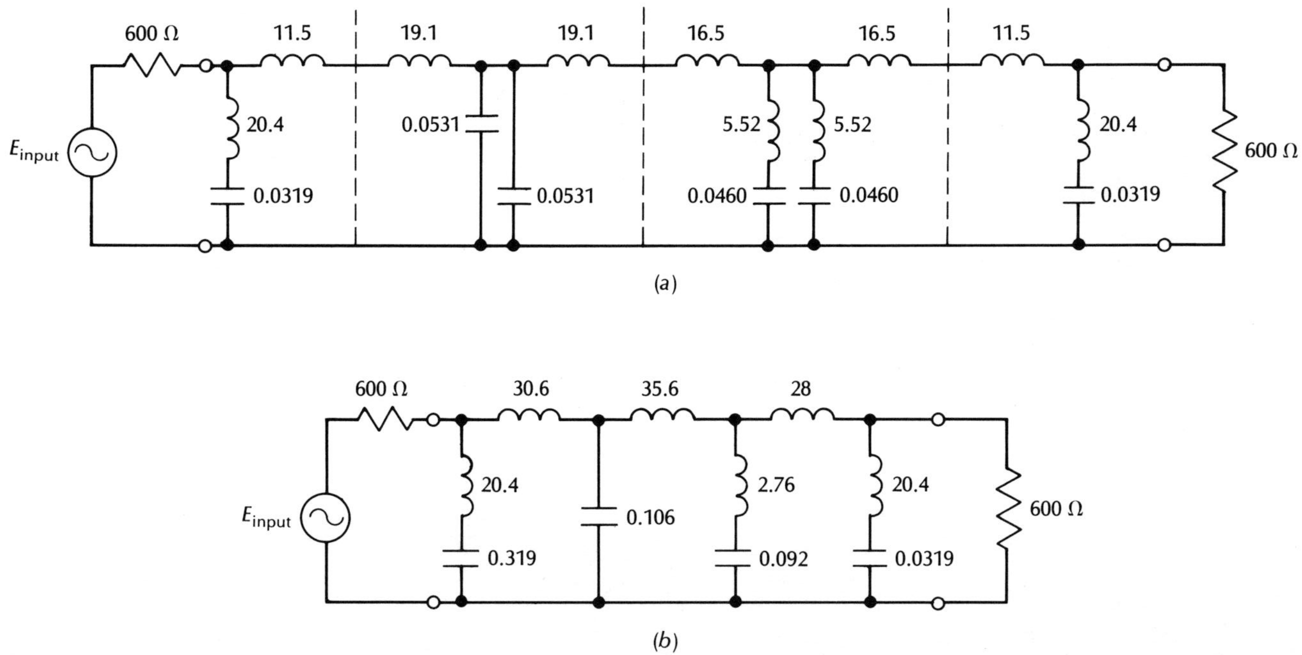

FIGURE 12-7 (*a*) Example filter with individual sections; (*b*) combined filter.

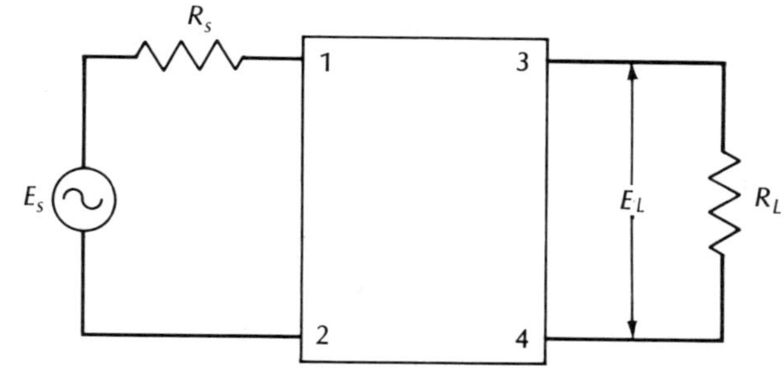

FIGURE 12-8 Generalized filter.

Modern network theory solves these equations to determine the network values for optimum performance in some respect.

The frequency response of the generalized filter can be expressed as a ratio of two polynomials in s, where $s = j\omega$ ($j = \sqrt{-1}$ and ω, the frequency in radians per second, is $2\pi f$) and is referred to as a *transfer function*. This can be stated mathematically as

$$T(s) = \frac{E_L}{E_s} = \frac{N(s)}{D(s)} \tag{12.20}$$

The roots of the denominator polynomial $D(s)$ are called *poles*, and the roots of the numerator polynomial $N(s)$ are referred to as *zeros*. Deriving a network's transfer function could become quite tedious and is beyond the scope of this book. However, the following discussion explores the evaluation and representation of a relatively simple transfer function. Figure 12-9 depicts a lowpass filter. Analysis of this filter results in the following transfer function:

$$T(s) = \frac{1}{s^3 + 2s^2 + 2s + 1} \tag{12.21}$$

The expression in (12.21) can be evaluated at different frequencies after substituting $j\omega$ for s. The result will be expressed as the absolute magnitude of $T(j\omega)$ and the relative attenuation in decibels with respect to the response at DC. Figure 12-9*b* shows the frequency-response curve. Analysis of equation (12.21) indicates that the denominator of the transfer function has three roots or poles and the numerator has none. The filter is therefore called an *all-pole type*. Since the denominator is a third-order polynomial, the filter is also said to have an $n = 3$ complexity. The denominator poles are $s = -1$, $s = -0.5 + j0.866$, and $s = -0.5 - j0.866$.

These complex numbers can be represented as symbols on a complex-number plane. The abscissa is α, the real component of the root, and the ordinate is β, the imaginary part. Each pole is represented as the symbol X, and a zero is represented as O. Figure 12-10 illustrates the complex-number plane representation for the roots of equation (12.21). There are certain mathematical restrictions on the location of poles and zeros in order for the

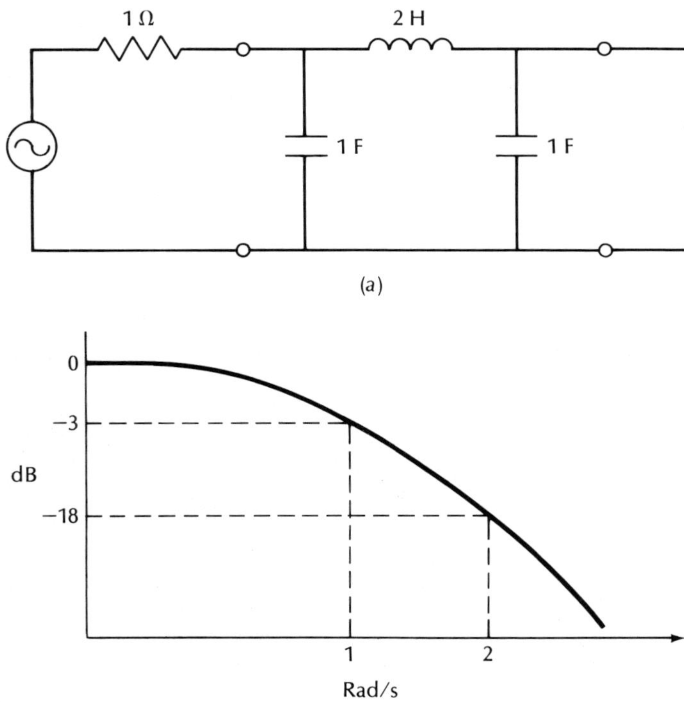

FIGURE 12-9 All-pole $N = 3$ lowpass filter circuit and frequency response.

$$T(j\omega) = \frac{1}{1 - 2\omega^2 + j(2\omega - \omega^3)}$$

ω	$\lvert +(j\omega)\rvert$	$20 \log \lvert t(j\omega)\rvert$, dB
0	1	0
1	0.707	−3
2	0.124	−18
3	0.037	−29
4	0.0156	−36

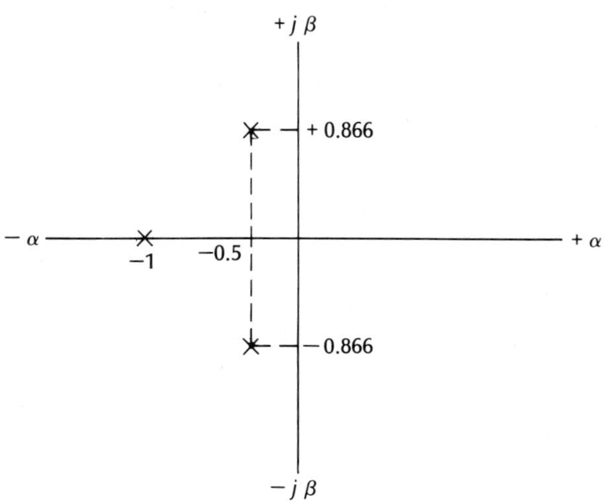

FIGURE 12-10 Complex-frequency plane representation of equation (12.21).

filter to be realizable. They must occur in pairs that are conjugates of each other, except for real-axis poles and zeros, which may occur singly. Poles must also be restricted to the left half-plane (the real coordinate of the pole must be negative). Zeros may occur in either plane.

Inverter Output Filter Considerations

Maintaining a high efficiency is very important in power conversion systems; therefore, the first requirement is that the filter elements be nondissipative. The filter must be made up of only reactive elements, namely, inductors and capacitors, with as low a dissipation as is possible in practical components. The most obvious function of the filter is that of waveshaping. The particular requirements imposed on the filter for a specific application will consist primarily of the specification of the maximum total harmonic content in the output voltage. This requires the filter to have a lowpass characteristic. The necessary attenuation of the particular filter will, of course, be as much a function of the waveshape of the voltage applied to the filter as it is of the desired quality of the output waveshape. In the case of a simple inverter whose output waveform is essentially a rectangular wave of voltage, the harmonic content of the voltage is illustrated in Fig. 12-11.

It is readily seen that if a sine-wave output voltage of low harmonic distortion is desired, a good lowpass or bandpass filter will be necessary to attenuate the lowest harmonic, the third. It must be further noted that only the lowest few harmonics must be considered if a filter with even moderate harmonic attenuation is used, since the total RMS harmonic voltage will be almost unaffected by higher harmonics. The ninth and higher harmonics can usually be ignored. The next desirable characteristic of a filter is that of independence of load variations. There are many types of waveshaping filters whose transfer functions vary widely with load variations. In order to maintain a constant output voltage, the input voltage to the inverter must necessarily be variable

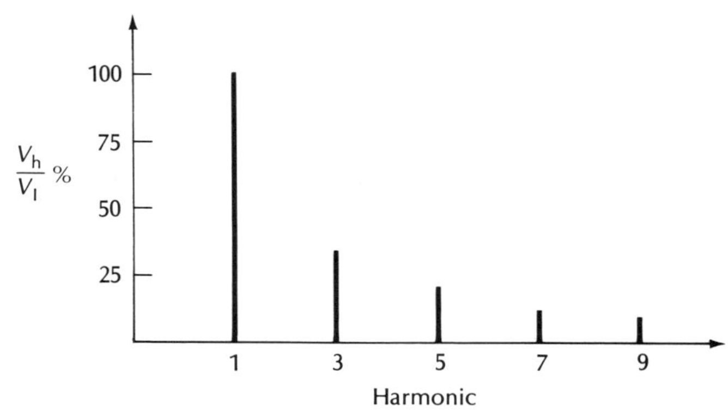

FIGURE 12-11 Harmonic content of inverter square wave.

over as large a variation as is encountered in the change in transfer function with load. An extremely desirable characteristic for the filter would be that of an insensitivity of the voltage-transfer function to variations in load.

So far, the characteristics imposed on the filter are those more of less normally required of a waveshaping filter. In addition, several other desirable features include minimum practical number, size, and weight of elements in the filter. Modern control circuit techniques generally produce a waveform into the output filter that is already sine-weighted, as opposed to a rectangular waveform, thus minimizing the filter requirements. Yet the simplest inverter type remains one that provides a rectangular output. The following synthesis of a filter circuit is provided for such an inverter.

Design Example

The filter network will be treated as a four-terminal network, as depicted in Fig. 12-12, for which the following equations apply:

$$I_1 = y_{11}E_1 + y_{12}E_2 \tag{12.22}$$

$$I_2 = y_{21}E_1 + y_{22}E_2 \tag{12.23}$$

where y_1, y_{11}, and y_{22} are the short-circuit admittances of a four-terminal network. It can be readily seen then that

$$I_2 = -E_2 y_L \tag{12.24}$$

where y_L is the load admittance. From equations (12.23) and (12.24) the voltage-transfer ratio can be shown that

$$\frac{E_2}{E_1} = \frac{-y_{12}}{y_{22} + y_L} \tag{12.25}$$

Equation (12.25) shows the voltage-transfer function in general to be a function of the load admittance.

The synthesis method employed in a synthesis of a two-terminal pair ladder network is based on the following two observations:

1. In a purely reactive ladder network, the poles of y_{12} are simple. Moreover, y_{22} and y_{11} possess these same poles.

2. Every zero of the transfer admittance y_{12} of a reactive ladder network must be either a zero of a shunt element or a pole of a series element.

Hence the method consists first of appropriately identifying the form of the y_{22} and y_{12} functions, and then synthesizing the susceptance y_{22} as a ladder network, but in such a way that the zeros of y_{12} are produced. It can be shown

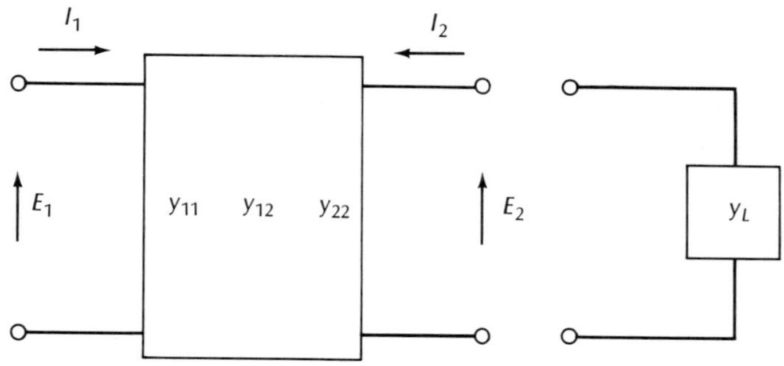

FIGURE 12-12 Short-circuit admittance characterization of a general four-terminal network.

that in order that a function be realizable as a reactance (or susceptance) it must be expressible as a polynomial. For reactive elements:

$$H(S) = \frac{KS(S^2 + \omega_2^2)(S^2 + \omega_4^2)\ldots}{(S^2 + \omega_1^2)(S^2 + \omega_3^2)\ldots} \tag{12.26}$$

where $\omega_1 < \omega_2 < \omega_3 < \omega_4 \ldots$.

Function y_{22} must be a realizable susceptance, and y_{22} and y_{12} must be identified. From equation (12.25) it is seen that if E_2/E_1 is to be independent of y_L at the fundamental frequency, then y_{12} and y_{22} must be $>>y_L$ at the fundamental frequency. Thus y_{12} and y_{22} must both possess poles at the fundamental frequency.

In order to determine the minimum number of elements necessary to satisfy all the desired requirements on the filter several cases will be considered. It should be noted that the number of elements in a given filter will be $N + 1$, where N is the number of critical frequencies (other than 0 or ∞) specified in y_{22} and y_{12}. Thus, a two-element filter specifies one critical frequency, and so on. The case of a two-element filter where E_2/E_1 is independent of load is the trivial case of the series LC combination in series with the load. In this case the series LC presents no impedance at the fundamental frequency and the transfer function is unity. The case of the three element filter with transfer function independent of load can be shown to be insufficient to satisfy all the desired conditions imposed on the filter. It does not appear that the three desirable characteristics (independence of load variation, capacitive input impedance, and sufficient harmonic attenuation) can be simultaneously satisfied in a three element filter. The case of the four element filter can be shown to be the filter with the minimum number of elements that still has a capacitive input impedance over a large range of load conditions and whose transfer function is independent of load. Thus for the four-element filter:

$$y_{22} = \frac{S(S^2 + \omega_2^2)}{(S^2 + \omega_1^2)(S^2 + \omega_3^2)} \tag{12.27}$$

$$y_{12} = \frac{KS}{(S^2 + \omega_1^2)(S^2 + \omega_3^2)} \tag{12.28}$$

Since both y_{22} and y_{12} must possess a pole at the fundamental frequency in order to render E_2/E_1 independent of load, ω_1 will be chosen to be unity (the fundamental frequency). The location of the remaining pole and zero of y_{22} is arbitrary as long as they be chosen at such frequencies that the desired harmonic attenuation is obtained. Some experience on the part of the filter designer will aid in the location of the remaining zero and pole. In general, however, the closer the remaining poles and zeros approach the fundamental frequency, the sharper the cutoff and the greater the harmonic attenuation.

$$\omega_2 = \sqrt{2} \qquad \text{and} \qquad \omega_3 = 2 \tag{12.29}$$

Then y_{22} and y_{12} are completely specified except for the constant-k in equation (12.28). The factors of equations (12.27) are multiplied out and successive synthetic divisions are performed to obtain the values of the components in henrys and farads for a 1-Ω load radian per second frequency. The synthetic division of y_{22} is shown below:

$$y_{22} = \frac{S(S^2 + 2)}{(S^2 + 1)(S^2 + 4)} = \frac{(S^3 + 2S)}{S^4 + 5S^2 + 4} \tag{12.30}$$

$$\begin{array}{llllll}
S^3 + 2S & \overline{S^4 + 5S^2 + 4} \,\lfloor S & & & & \\
 & \underline{S^4 + 2S^2} & & & & \\
 & \quad 3S^2 + 4 & \overline{\lceil S^3 + 2S} \,\lfloor \frac{S}{3} & & & \\
 & & \underline{S^3 + \frac{4}{3}S} & & & \\
 & & \frac{2}{3}S & \overline{\lceil 3S^2 + 4} \,\lfloor \frac{9}{2}S & & \\
 & & & \underline{3S^2} & & \\
 & & & \quad 4 & \overline{\lceil \frac{2}{3}S} \,\lfloor \frac{S}{6} &
\end{array}$$

The synthesis of y_{22} can then be shown to be the circuit of Fig. 12-13. By open-circuiting the input terminals the network becomes the two-terminal pair filter network desired. The filter design may now be reduced to a specific frequency and power level by multiplying all inductors by Z_d/ω_d, and all capacitors by $1/Z_d\omega_d$ where Z_d is the desired design impedance level of the filter

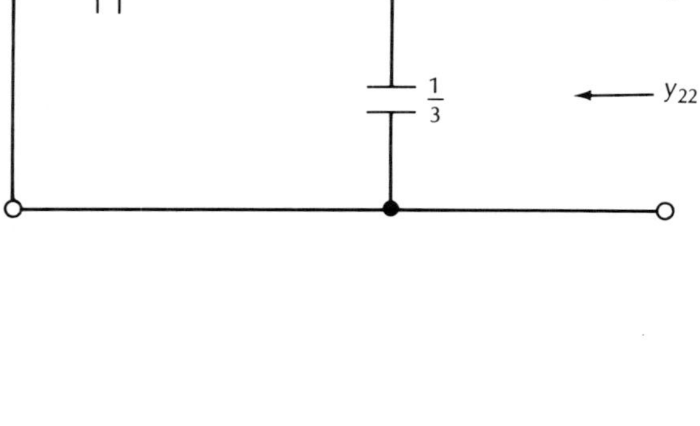

FIGURE 12-13 Realization of the y_{22} admittance.

and ω_d is the design frequency. The constant K can be evaluated by examining the network and equation (12.28) at $S \to 0$:

$$\text{For equation (12.28):} \quad y_{12}|_{S\to 0} \rightrightarrows \frac{KS}{4} \tag{12.31}$$

$$\text{For network:} \quad y_{12}|_{S\to 0} \rightrightarrows \frac{S}{6} \tag{12.32}$$

Therefore, for the network and the y_{12} function to be equivalent $ks/4 = S/6$ and $k = 2/3$. Equation (12.25) becomes

$$\frac{E_2}{E_1} = \frac{\frac{2}{3}S}{S(S^2 + 2) + y_L(S^2 + 1)(S^2 + 4)} \tag{12.33}$$

from which it is seen that E_2/E_1 is ideally independent of load at the fundamental frequency. It is also readily seen that the filter has a bandpass characteristic. All that remains is to show that the filter has a capacitive input impedance. It is possible to calculate Z_{input} from the network analysis on the filter circuit itself from which

$$Z_{\text{in}} = \frac{j9Z_L}{-2Z_L + j4} \tag{12.34}$$

where Z_L = per unit load impedance.

It should be noted that the no-load impedance is 2.25 times the full-load impedance and is completely reactive. The input filter current magnitude for the no-load condition is less than half the full load current, in contrast to many conventional filters that take full-load current or more at the no-load condition. Also, note from equation (12.3 4) that the short-circuit current is theoretically infinite, which is a desirable characteristic for providing fuse clearing fault currents to loads that fail, without disturbing other loads on the same output.

CHAPTER 13
Feedback and Control Circuitry

In this chapter we will be dealing with output voltage regulation, feedback, and various control circuits useful for generating the desired waveforms. As mentioned earlier, there are times when a rectangular output waveform is satisfactory for a particular type of load. Often it is just as well with this type of waveform to allow the RMS voltage to vary, as the battery voltage varies, rather than add the cost and complexity of creating a closed-loop feedback system in an attempt to regulate the output. However, using a ferroresonant transformer in an inverter circuit will provide one possible solution of maintaining some degree of output voltage regulation, while keeping it simple. The price to pay for this simplicity, though, is size and weight, as discussed earlier in the section on voltage regulators.

Over the years there have been many clever methods employed to solve the problem of converting a DC voltage (battery) to an AC voltage. Smaller and faster power devices used in DC-to-AC inverters have evolved, and their implementation has been simplified. Until the early 1970s, power bipolar transistors were not available at a cost that would allow them to displace SCRs. Hot on the heels of bipolar advances came power MOSFET technology, and others, making it easier all the time for an engineer to implement a cost effective design. Just as power devices have advanced, so have digital and linear integrated circuits. ICs are getting faster, containing more functions per given size package, and consuming less power. An entire industry of switching power supply designers have attracted the attention of device manufacturers to the extent that they now have dedicated control ICs, with all the technical assistance and applications literature they need to help in the ultimate design of the end product. Much of this same technology is useful in DC-to-AC conversion as well, and should be utilized as much as possible. The typical approach to feedback regulation of power converters is to implement an operational amplifier to generate an error signal that ultimately provides the information necessary for the control circuit to maintain the proper output. If the amplifier is not properly applied to the circuit it may oscillate, causing a failure. It may not oscillate all the time, but just when certain loads are applied. The following pages will deal with the basic concepts of feedback and frequency compensation of operational amplifiers, followed by a discussion of various control techniques.

Basic Concepts of Feedback

There are many different types of feedback circuit. In the standard inverting configuration of Fig. 13-1*a*, negative feedback is obtained by connecting resistor R_f from the output to the input. The actual comparison of output and input takes place by combining currents i_1 and i_2 at the amplifier input node, hence the name "current differencing." In order to facilitate the analysis of this type of circuit the feedback resistor can be replaced by the equivalent circuit, as shown in Fig. 13-1*b*. If the feedback network is more complicated than the single resistor of Fig. 13-1*a*, then the equivalent circuit of Fig. 13-1*c* is used since any linear circuit can be put in this form. The effect of the $K_2 v_d$ source is negligible in most applications (in Fig. 13-1*a* $K_2 v_d = v_d \approx 0$ since the operational-amplifier input is a virtual short circuit), and so it is omitted from the final circuit of Fig. 13-2. This circuit is an example of a current-differencing negative-feedback amplifier, all of which consist conceptually of three sections:

1. An amplifier to which feedback is applied.
2. A feedback network which may contain anything from a single resistor to a circuit containing nonlinear elements.
3. A differencing, or summing, circuit in which the output is compared with the input. This comparison is the essence of feedback.

The circuit of Fig. 13-3 differs from that of Fig. 13-2 in that the output of the feedback network is in series with the amplifier input resistance R_i in Fig. 13-3, while in Fig. 13-2 it is in parallel with R_i. The feedback illustrated in Fig. 13-3 is representative of negative feedback with voltage differencing. The comparison of output and input takes place via the voltages Kv_L and v_i in a series loop. The gain of the amplifier in Fig. 13-2 is calculated by first expressing the currents at the amplifier input node in the form

$$i_1 = \frac{v_i + v_2}{R_s} \qquad i_2 = \frac{v_d + Kv_L}{R_f} \tag{13.1}$$

which leads to

$$i_d = i_1 - i_2 = \frac{v_i + v_d}{R_s} + \frac{v_d + Kv_L}{R_f} = -\frac{v_d}{R_i} \tag{13.2}$$

Solving for v_d in (13.2), assuming that $R_i >> R_s$ or R_f, and $R_\beta >> R_o$ so that $v_L \approx A_d v_d$, we find, after some manipulation:

$$v_L \approx -\frac{(A_d v_i/R_s)(R_s \| R_f)}{1 + (A_d K/R_f)(R_s \| R_f)} \tag{13.3}$$

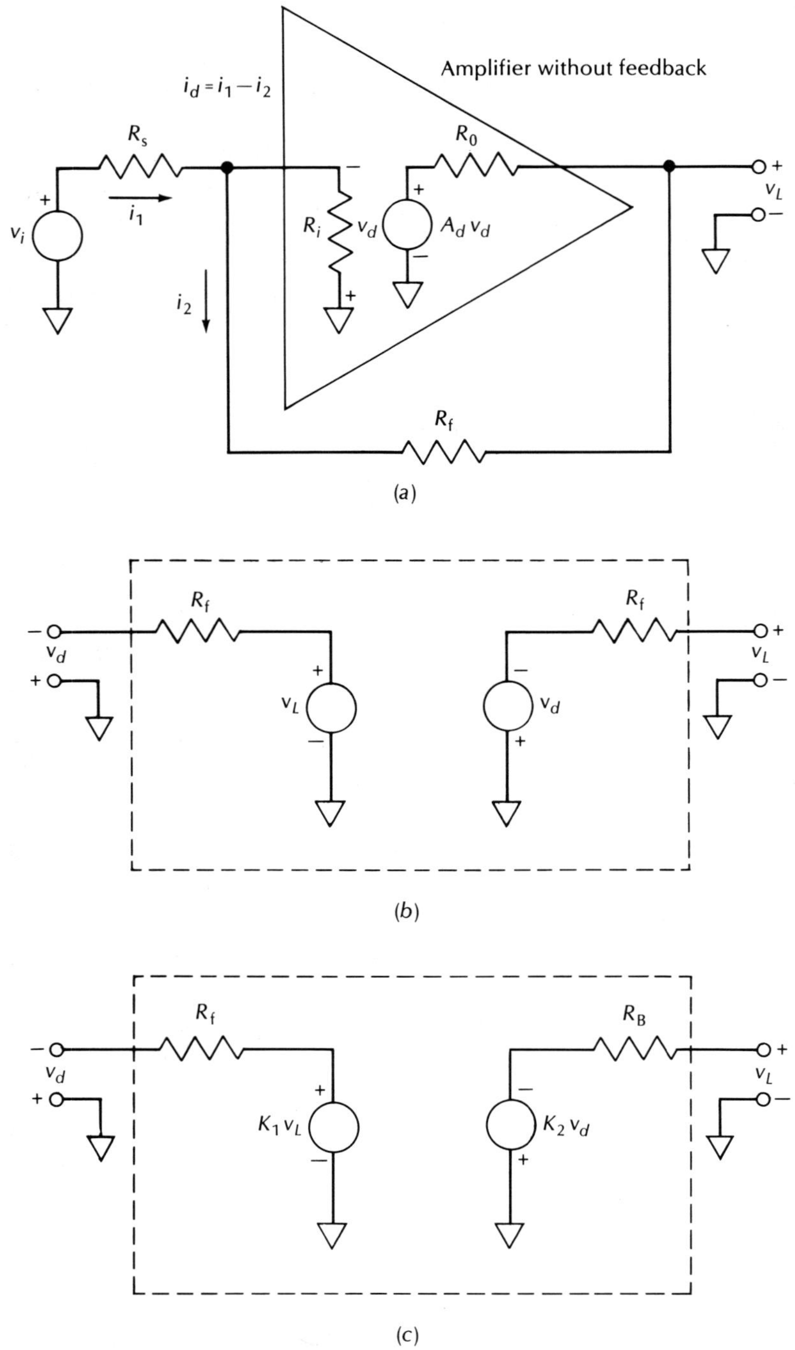

FIGURE 13-1 Feedback circuits: (*a*) inverting configuration; (*b*) feedback network equivalent circuit; (*c*) general equivalent circuit.

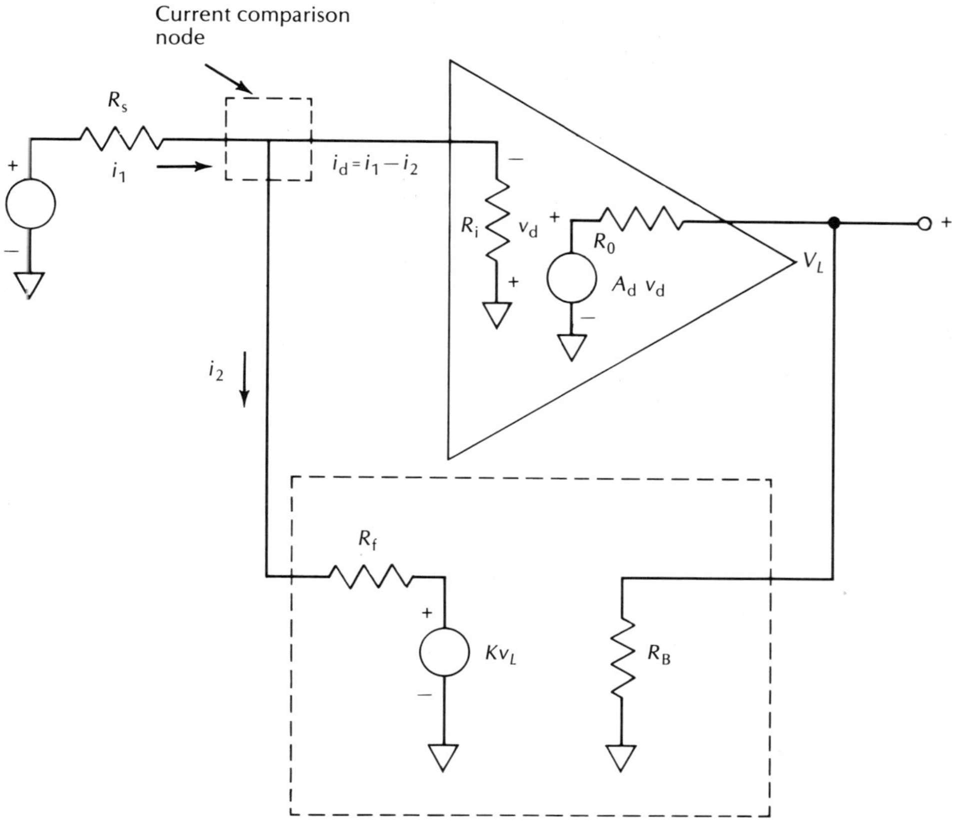

FIGURE 13-2 Current-differencing negative-feedback circuit.

when the term in the denominator is much greater than unity, then

$$\frac{A_d K}{R_f}(R_s \| R_f) >> 1 \tag{13.4}$$

Equation (13.3) will reduce to

$$v_L \approx -\frac{R_f}{KR_s} v_i \tag{13.5}$$

which is independent of the amplifier gain A_d. For the standard inverting configuration of Fig. 13-1*a*, $K = 1$, so that $v_L = -(R_f/R_s)v_i$. The loop gain T of the amplifier is the term found in the denominator of (13.3). This gain can be calculated directly using the circuit of Fig. 13-4, which differs from that of Fig. 13-2 in that the voltage source in the feedback network is called Kv_L' rather than Kv_L. The loop gain is defined as the gain around the circuit

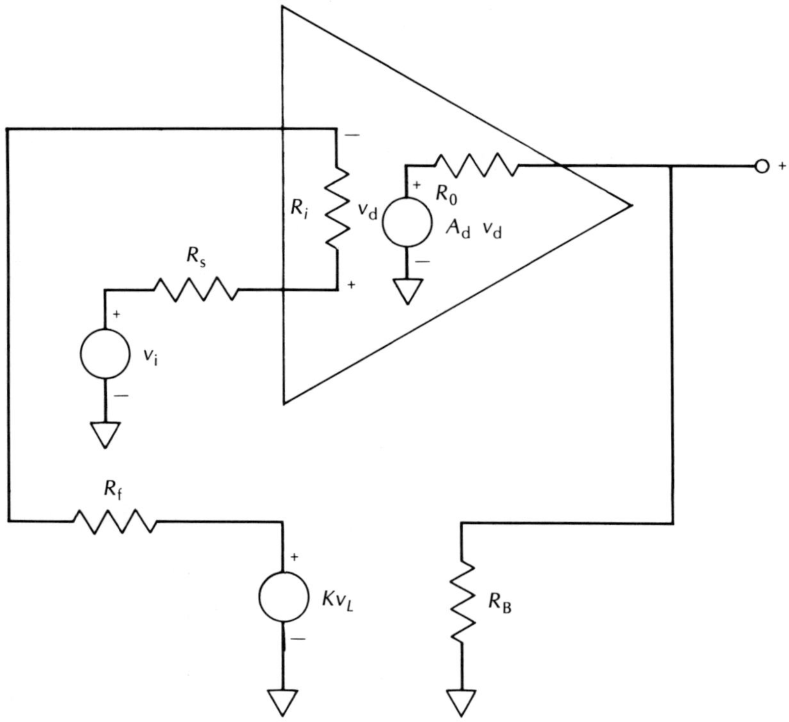

FIGURE 13-3 Voltage-differencing negative feedback circuit.

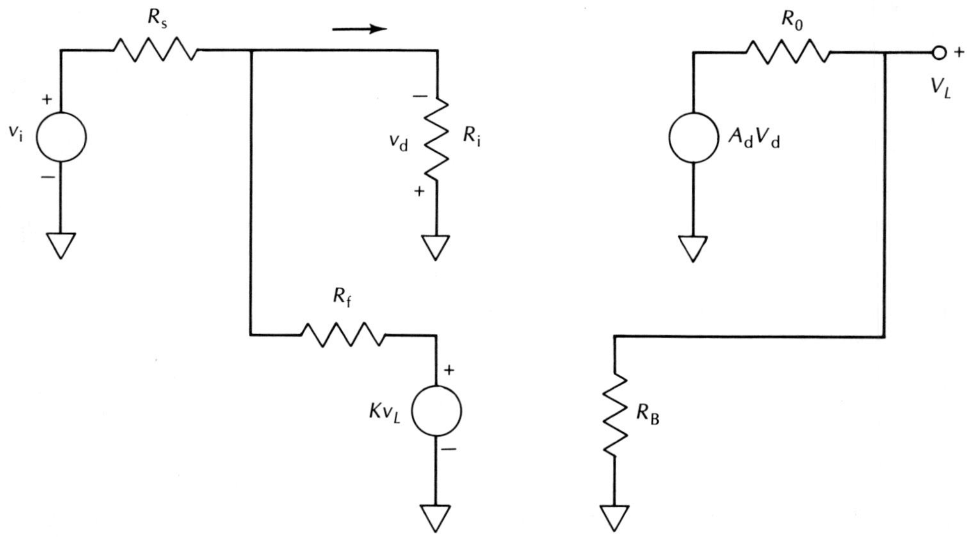

FIGURE 13-4 Circuit used to calculate loop gain t and the gain without feedback A_0.

loop with the input voltage equal to zero:

$$T = \left.\frac{v_L}{v'_L}\right|_{v_i=0} = \frac{v_L}{v_d}\frac{v_d}{i_d}\frac{i_d}{v'_L}$$
$$T \approx A_d(-R_i)\left(\frac{K}{R_s + R_f}\right)\left(\frac{R_s}{R_s + R_i}\right) \tag{13.6}$$

Assuming that $R_i >> R_s$ or R_f, and $R_\beta >> R_o$, we have

$$T \approx -A_dK\frac{R_s}{R_s + R_f} = -\frac{A_dK}{R_f}(R_s \| R_f) \tag{13.7}$$

The numerator of (13.3) is called the *open-loop gain* or the *gain without feedback* A_0. This gain can be found using the circuit of Fig. 13-4 by setting $v'_L = 0$ in the following:

$$A_o = \left.\frac{v_L}{v_i}\right|_{v'_L=0} = \frac{v_L}{v_d}\frac{v_d}{v_i} \approx A_d\frac{-R_f}{R_s + R_f} = -\frac{A_d}{R_s}(R_s \| R_f) \tag{13.8}$$

Using the results of (13.7) and (13.8), we can rewrite (13.3) as

$$A_v = \frac{v_L}{v_i} = \frac{A_0}{1 - T} \tag{13.9}$$

This expression gives the overall gain A_v in terms of the gain without feedback A_0 and the loop gain T for the current-difference circuit of Fig. 13-2. The gain of the feedback amplifier shown in Fig. 13-3 can also be found using (13.9) by changing the feedback voltage source Kv_L to $K'v_L$ and going through the numbers, using the same assumptions as above.

From (13.9) it is shown that the loop gain determines the amount of feedback present in a circuit. When $T = 0$ ($K = 0$), there is no feedback. When $-T$ becomes very large compared with unity, the gain of the feedback amplifier approaches $A_v = v_L/v_i = -(R_f/KR_s)$ for the circuit shown in Fig. 13-2, and $A_v = 1/K$ for the circuit of Fig. 13-3. In either case the gain with feedback is approximately independent of the gain of the amplifier, A_d. The gain with feedback becomes more immune to the amplifier temperature and variations in parameters as the loop gain increases. This is an important benefit of feedback.

So far it has been assumed that A_0 and T are independent of frequency and negative feedback is employed so that T is a negative number and $1 - T$ is a positive number. In an amplifier, response does depend on frequency, so that while T may be negative at low frequencies, its phase will generally increase at high frequencies. In order to determine the frequency response of a typical feedback amplifier, refer to the circuit of Fig. 13-2. The overall gain for this circuit is

$$A_v = \frac{A_0}{1 - T} \tag{13.10}$$

where A_0 and T are assumed to be constants. Set that assumption aside now, and let A_0 and T be functions of the complex frequency s.

Considerations for a Single-Pole Amplifier

Assume that the amplifier gain A_d has a single negative real pole at $s = \omega_1$, so that

$$A_d(s) = \frac{A_{dm}}{1 + s/\omega_1} \tag{13.11}$$

where A_{dm} is the gain at low frequencies. If the feedback network is frequency independent, the loop gain T (13.7) can be written as

$$T = -\left(\frac{A_{dm}K(R_s \| R_f)}{R_f}\right)\frac{1}{1 + S/\omega_1} \tag{13.12}$$

If we define T_m as the magnitude of the loop gain at frequencies below ω_1, then

$$T_m = -\frac{A_{dm}K(R_s \| R_f)}{R_f} \tag{13.13}$$

and

$$T = \frac{T_m}{1 + s/\omega_1} \tag{13.14}$$

The open-loop gain A_o will be (13.8):

$$A_o = \frac{A_{om}}{1 + s/\omega_1} \tag{13.15}$$

where A_{om}, the magnitude of the open-loop gain at low frequencies, is

$$A_{om} = \frac{A_{dm}(R_s \| R_f)}{R_s}$$

$$A_v = \frac{v_L}{v_i} = \frac{-A_{om}/(1 + s/\omega_1)}{1 + T_m/(1 + s/\omega_1)} \tag{13.16}$$

Substituting (13.14) and (13.15) into (13.9) will give

$$A_v = \frac{-A_{om}}{1 + T_m}\left(\frac{1}{1 + s/\omega_1(1 + T_m)}\right) \tag{13.17}$$

From (13.17) it is shown that the feedback amplifier has a pole at

$$s = -\omega_1(1 + T_m) \tag{13.18}$$

and therefore the upper 3-dB frequency of the amplifier with feedback is

$$f_h = f_1(1 + T_m) \tag{13.19}$$

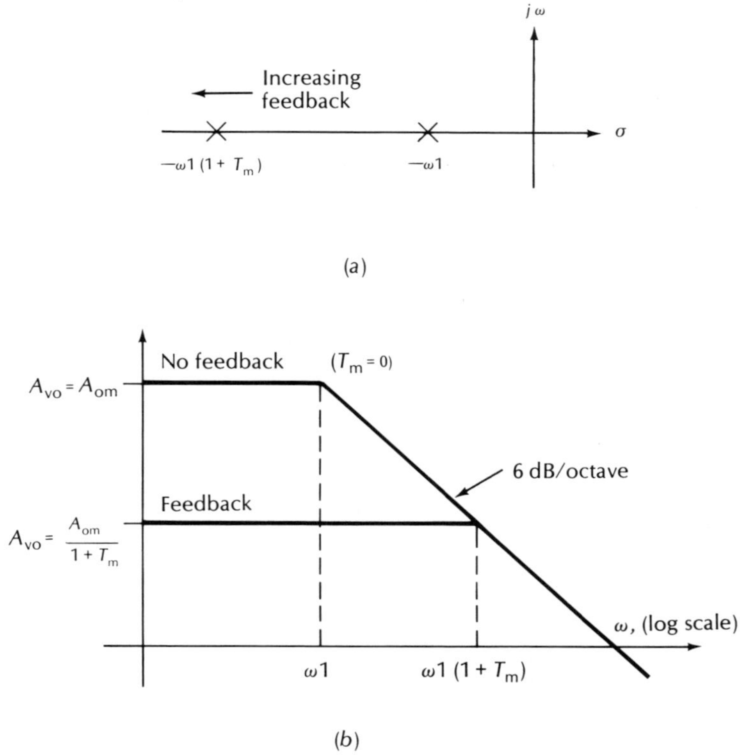

FIGURE 13-5 Characteristic of single-pole feedback amplifier: (*a*) locus of pole motion; (*b*) gain versus frequency.

So the 3-dB frequency is increased by the loop gain while the midfrequency gain is decreased by the same amount. The gain bandwidth product *GBW* is then

$$GBW = [A_v(f = 0)]\, f_h = A_{om} f_1 \tag{13.20}$$

which is a constant, independent of the feedback.

Without feedback, the pole is at $-\omega_1$, and as the amount of feedback is increased from zero, the pole moves farther into the left half-plane along the negative real axis, as shown in Fig. 13-5*a*. This is valid only if *T* is a negative number. If *T* is a positive number, the pole will move into the right half-plane and the amplifier will become unstable. The root locus curve of Fig. 13-5*b* shows how the poles of the overall transfer function vary in terms of the variation of a parameter, in this case, the loop gain.

Considerations for a Double-Pole Amplifier

A more complicated case is assumed now: an amplifier gain function with a pair of coincident real poles.

$$A_d(s) = \frac{A_{dm}}{(1 + s/\omega_1)^2} \tag{13.21}$$

Equation (13.12) is now written as

$$T = -\frac{A_{dm}K(R_s \| R_f)}{R_f}\left(\frac{1}{1 + s/\omega_1}\right)^2 = \frac{T_m}{(1 + s/\omega_1)^2} \tag{13.22}$$

and (13.15) becomes

$$A_0 = -\frac{A_{om}}{(1 + s/\omega_1)^2} \tag{13.23}$$

and finally, (13.10) becomes

$$A_v(s) = \frac{-A_{om}/(1 + s/\omega_1)^2}{1 + T_m/(1 + s/\omega_1)^2} = \frac{-A_{om}}{1 + T_m + 2s/\omega_1 + s^2/\omega_1^2} \tag{13.24}$$

Applying a little algebra, we can put (13.24) in the standard form of

$$A_v(s) = \frac{-A_{om}}{1 + T_m}\left(\frac{1}{1 + 2\zeta s/\omega_n + s^2/\omega_n^2}\right) \tag{13.25}$$

where

$$\omega_n = \omega_1\sqrt{1 + T_m} \tag{13.26}$$

and

$$\zeta = \frac{1}{\sqrt{1 + T_m}} \tag{13.27}$$

The voltage gain given in (13.25) shows that the system is of second order, with poles at

$$s = -\omega_n(\zeta \pm \sqrt{\zeta^2 - 1}) \tag{13.28}$$

The poles of the overall transfer function are negative, real, and coincident when there is no feedback, but become complex as soon as $T_m > 0$. The loci of these poles for increasing T_m are simply two vertical straight lines, as shown in Fig. 13-6*a*. The asymptotic and actual frequency responses with and without feedback are shown in Fig. 13-6*b*. Since the poles of $A_v(S)$ are in the left half-plane for all values of feedback, there is no stability problem in the strict sense. When the feedback is large, however, the frequency response exhibits a sharp peak. This leads to damped oscillations in the response of the circuit, which are usually undesirable. The magnitude of the peak depends on the damping ratio and increases with increasing feedback. Normalized plots of the magnitude and phase of the frequency response are shown in Fig. 13-7 for various values of damping ratio.

The assumptions and observations made here for single- and double-pole amplifier considerations can be carried on into consideration of stability problems associated with three-pole amplifiers as well. While there are many good books dealing specifically with this topic, the purpose of this discussion was

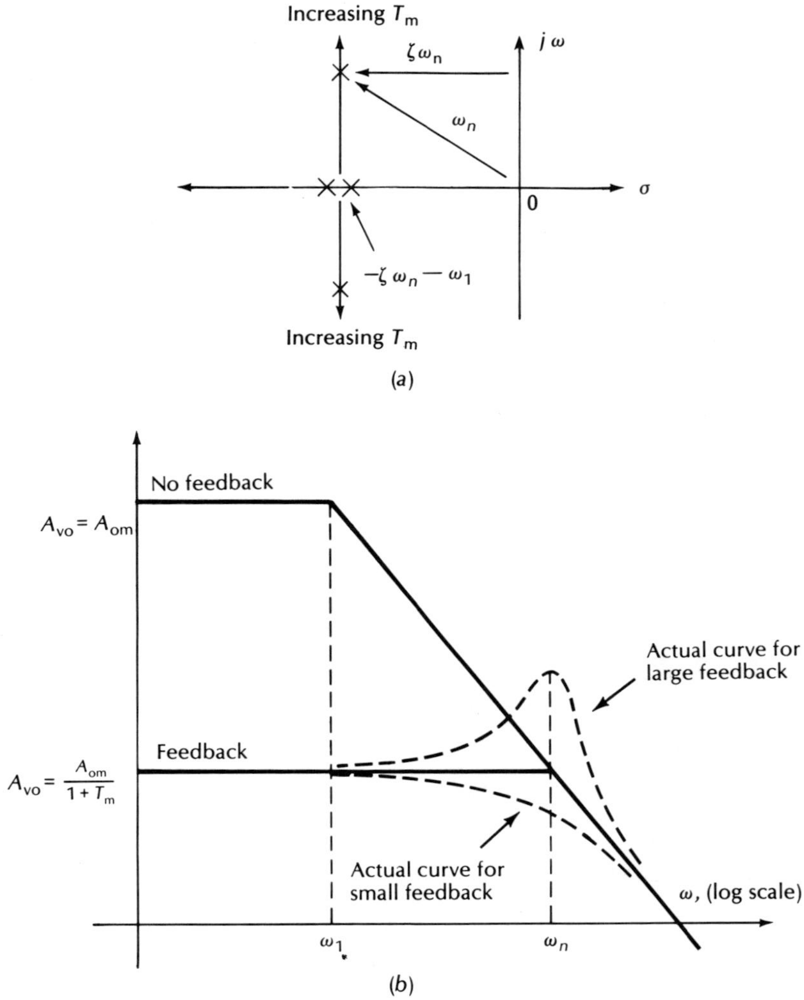

FIGURE 13-6 Characteristic of two-pole feedback amplifier: (*a*) root locus; (*b*) gain versus frequency.

to refresh the reader on some of the considerations necessary to applying amplifiers in feedback networks, and point out to those new to the subject a few of the precautions that lead to stable feedback systems. With these considerations in mind we will now describe some applications of control circuits in switching power converters.

Inverter Control Circuitry

As it was stated earlier, not all DC to AC inverters utilize any form of output voltage feedback or regulation. These inverters, usually possessing a rectan-

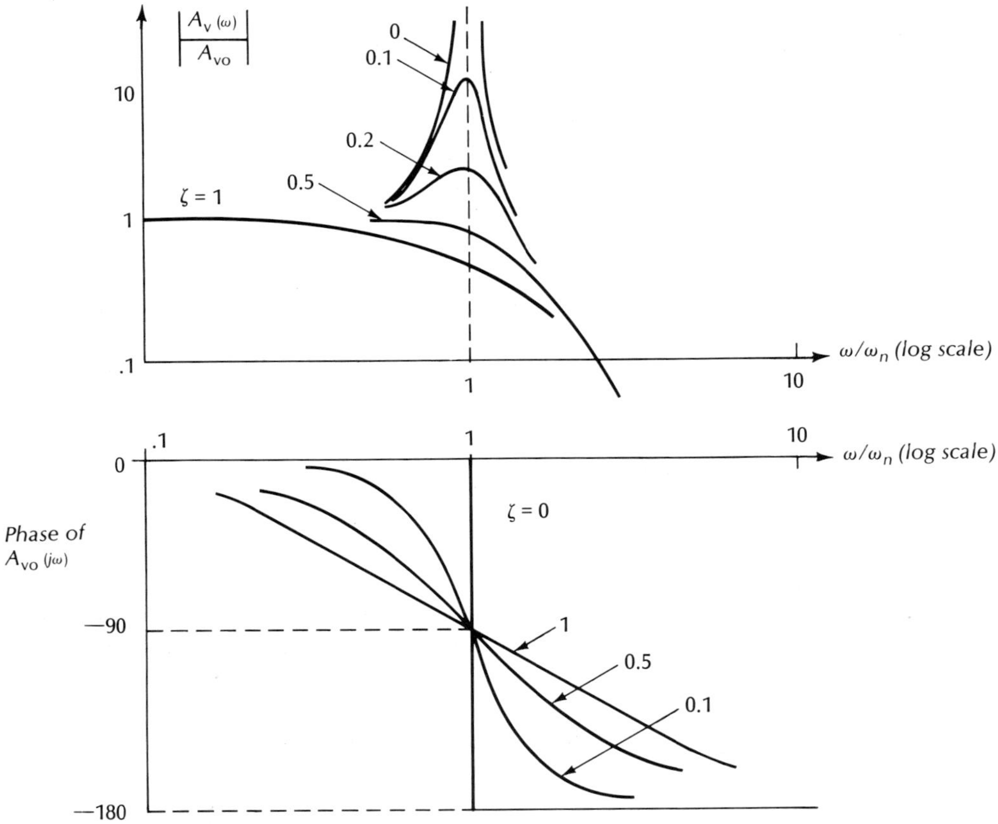

FIGURE 13-7 Response of second-order system.

gular (square wave) output waveform, have been around for a long time, and have found a niche in the marketplace. The lower cost associated with the generation of a nonregulated rectangular output voltage waveform has led to the development of both online uninterruptible power systems and standby power systems intended to serve the microcomputer market. Typical control circuitry used for generating rectangular or quasi-rectangular, nonregulated outputs is illustrated in Fig. 13-8.

The primary difference between the full square-wave type of output and the stepped or quasi-square is the ability to more closely approximate a sine wave. A quasi-square waveform, with the correct dwell or dead time set, will have reduced harmonic content over that of a full square wave. Figure 13-9 depicts the relationship between the dwell angle and the harmonic voltage content (3rd, 5th, 7th) of the output waveform. As can be seen, there is no magic dwell angle that will make the harmonics disappear. But there are angles that will allow the minimization of one or more of the major harmonics, 3, 5, and 7. By setting the output waveform to be on for 120°, out of the 180°

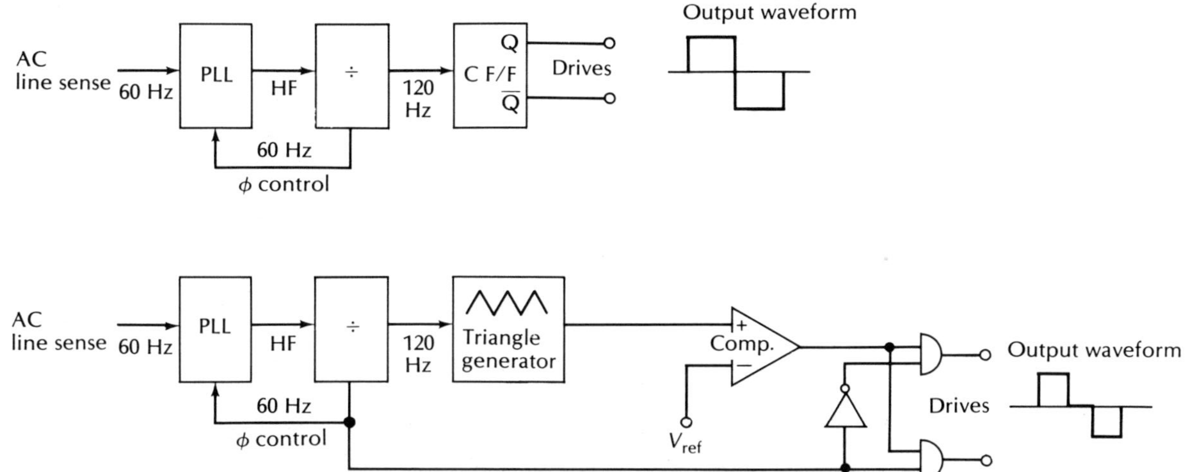

FIGURE 13-8 Simple control circuits providing square- and quasi-square-wave drives, phase-locked to the power line. Dwell angle of the quasi-square-wave drive is set by V_{REF}; for regulation V_{REF} can be adjusted by a feedback amplifier (PLL = phase-locked loop).

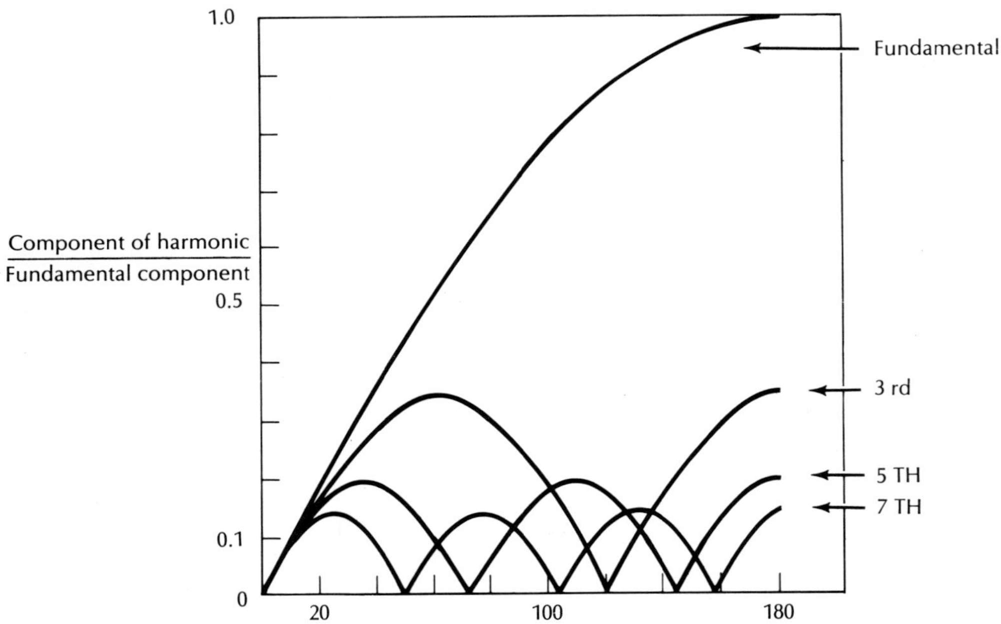

FIGURE 13-9 Content of third, fifth, and seventh harmonic as a percentage of the fundamental frequency. Note that at a full square wave, (180°), all three harmonics are at maximum, while at 120° the third is at zero.

half-cycle, the third is reduced. That in itself is a big help with most loads that care about waveshape, such as motors.

One of the prime reasons for not regulating the quasi-square output waveform is to maintain the 120° output. Typically the output voltage for this waveform would be regulated by adjusting the width as the battery drains, maintaining a constant RMS voltage at the output. A method for doing this is shown in Fig. 13-10.

The main difficulty with regulating this waveform is that the dwell angle, which has been shown to be best at 120°, must now change in order to maintain a constant RMS output. Another method of adjusting this waveform, without the need for feeding output voltage back to the control circuit is to monitor the battery voltage and adjust the output waveform as the DC voltage varies. This method will usually yield acceptable performance, although it is not as accurate as using a full output voltage feedback method.

One major complaint with the full square-wave output is that, for example, in order to maintain enough output voltage to operate the switching power supply on the input of a computer at a low battery condition, the peak output voltage at a high battery condition must be very high. Full battery voltage on a lightly loaded square wave output inverter can result in output voltages in excess of 160 V RMS. Since the peak voltage is the same, and the peak voltage of a 120-V RMS sine wave is ≈ 170 V, many feel this is not a problem. But

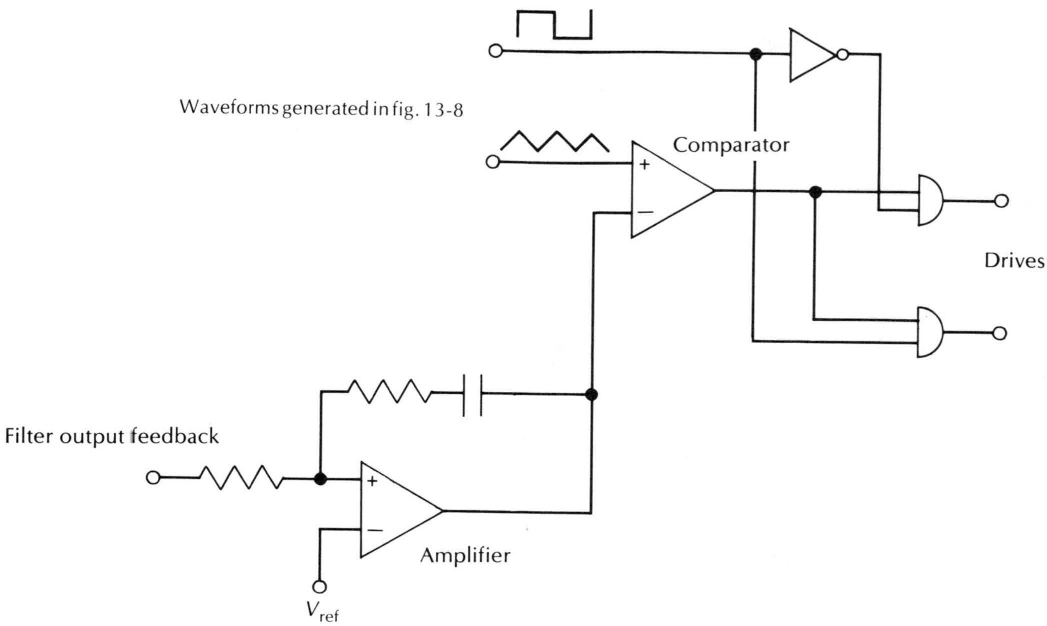

FIGURE 13-10 Control circuit of Fig. 13-8, modified to add feedback for output regulation.

the sine wave is at 170 V for only a very small portion of the period. The square wave will be high for the entire period. This could result in overheating in some components of the load and does need to be considered.

If it is desired to maintain the cost-effectiveness of the square-wave output, without adding regulation as in the quasi-square output, there is another method that works quite well. If the output transformer for the inverter, as shown in Fig. 13-11, were designed to saturate when the battery voltage is above some number, then a current limit circuit would terminate the drives to the power transistors at a predesignated point. Using current, rather than voltage, information to regulate the output is just another form of feedback. Since the only time the high output is a problem with this approach is when the battery is high, either adding more load, which discharges the battery quickly below the saturation point, or having the battery discharge slowly into a light load will cause the transformer to operate in saturation for only a portion of the time. This scheme works best in a standby inverter where it is not required to operate continuously. In this circuit the current limit is determined by measuring the V_{BE} on the power transistors. With bipolar devices this voltage will rise as the transistor begins to pull out of saturation due to excessive collector current. By measuring this base voltage and comparing it to a reference that is established as a voltage equivalent to the desired current limit level, the drives can be terminated for the remainder of the output half-

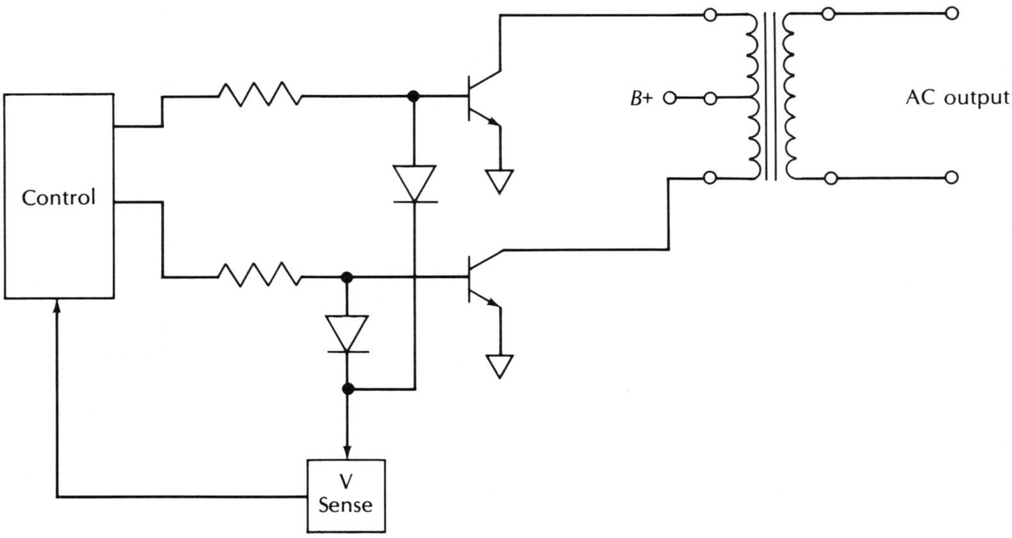

FIGURE 13-11 Square-wave drive modified to terminate on excessive V_{BE}.

cycle when that current is reached. The resultant waveform will be quasi-square at high battery, widening out to fully square as the transformer comes out of saturation at lower battery voltages.

The control circuitry necessary for generating a sine-wave output is really not too much more complex than that for generating the single-pulse-width-modulated waveform shown in Fig. 13-10. In fact, the single pulse control and its resultant waveform will provide a sine wave, just as a full square wave can be made to yield a sine wave, if you are willing to pay the cost, size, and weight for the output filter required. The whole purpose for increasing the complexity of the control circuit is to reduce the size of the output filter. One method is a variation on the circuit of Fig. 13-10. In order to get a single pulse in each half-cycle, a triangle waveform is compared to a DC voltage. In order to regulate the output of that circuit, the DC voltage can be moved up or down as required based on feedback information. To make more pulses per half-cycle, all that is required is to increase the frequency of the triangle as shown in Fig. 13-12*a*. Now, as the DC voltage is moved up and down each of the pulses will vary in width, changing the output voltage, for regulation as required.

If the DC voltage in Fig. 13-12*a* were modulated to move up and down in a sinusoidal fashion, at the same frequency as is desired at the output as shown in Fig. 13-12*b*, then the pulses generated would possess "sine-weighting," which is much easier to filter to the desired sine wave output, compared to a non-sine-weighted pulse string.

Just as with the square and quasi-square output circuits shown above, the generation of a sine-wave output in this case does not necessarily imply regu-

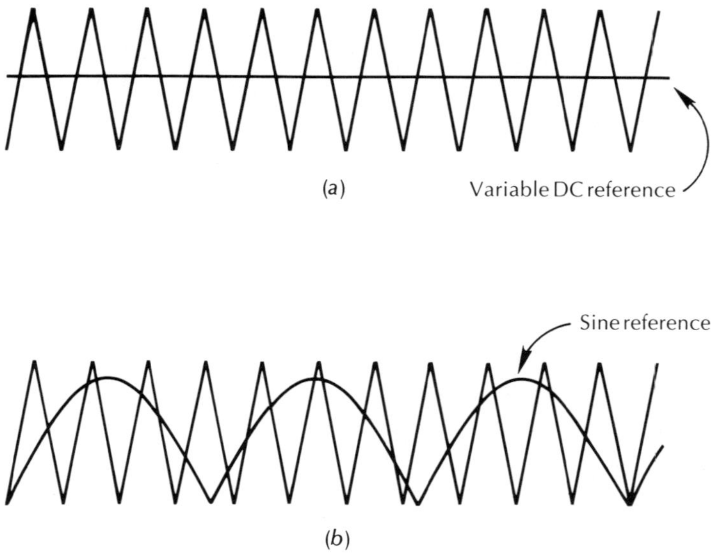

FIGURE 13-12 Illustration of how high-frequency triangle can be compared to a DC reference to generate variable-width pulses at triangle frequency. Modulating reference with a sine wave will allow generation of sine-weighted pulses.

lation. This "programmed waveform" technique is useful where the harmonic voltage content must be reduced, but large bulky filters are not desirable, and regulation is not of concern. Should it be desirable to regulate the output, then using an error amplifier, as shown in Fig. 13-13, to compare the output voltage feedback to some reference will work quite well. The output of the error amplifier is used to modify either the height of the triangle waveform, or the height of the sine reference waveform. Changing either will result in changing the output pulse width as necessary to regulate the output voltage.

The frequency of the pulse-width modulation (PWM) in this circuit is not too critical. The type of power devices used will be important in the decision. For example, the switching losses associated with slower bipolar transistors is good reason to keep the number of switching operations per output period down as much as possible, to minimize losses. Using power MOSFET devices will improve on this situation, but experience has shown there is not too much benefit to be realized, in the way of output waveform improvement, by operating the PWM frequency above 4 kHz. Audio frequency operation is a consideration, though, and if the losses can be kept reasonable, that alone may be cause to move to higher frequencies.

Another variation on this same type of control circuit, using sine weighted PWM, is achieved with the circuit shown in Fig. 13-14. Just as in the programmed waveform approach discussed above, the output sine wave is made up of many narrow pulses, whose width is determined by the feedback and regulation circuit. The difference is in the pulse-to-pulse relationship. In the programed waveform each pulse was set to a predetermined width by the control circuit, as necessary to create the sine-weighted PWM. Here, that is

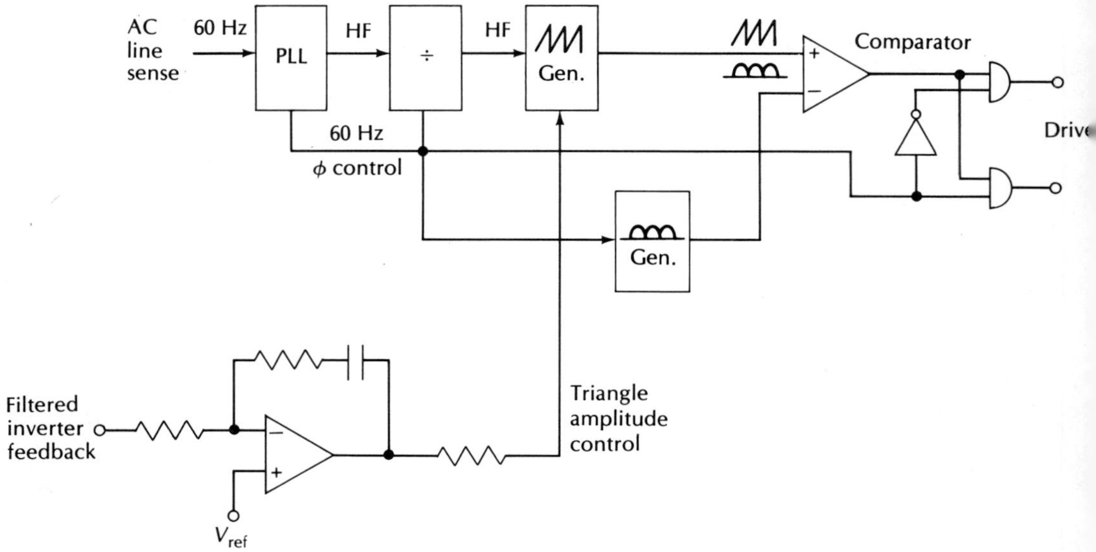

FIGURE 13-13 Sine-weighted PWM control with voltage feedback for regulation.

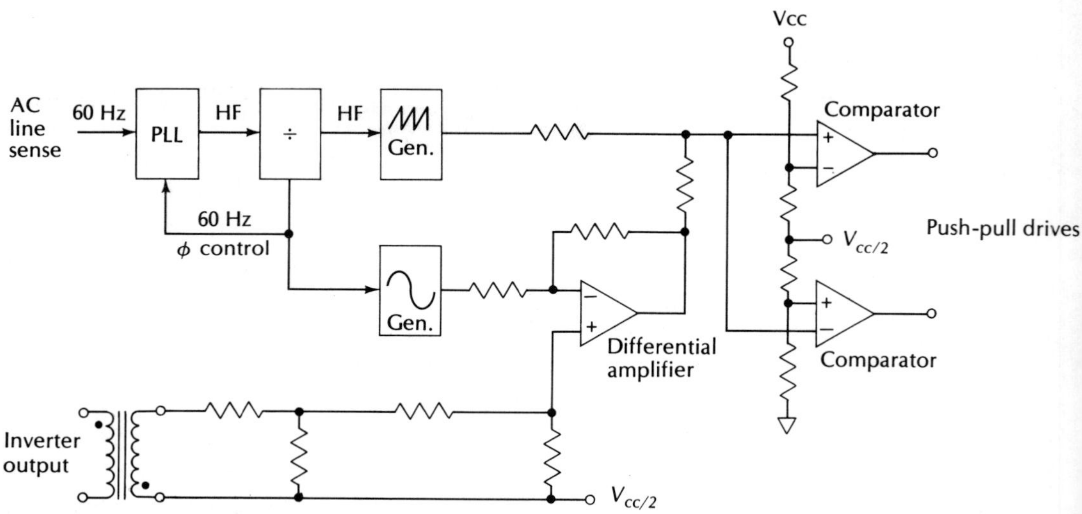

FIGURE 13-14 Push–pull control circuit using high-frequency PWM and fast regulation voltage control.

not the case. In the circuit of Fig. 13-14 a differential amplifier compares a reference sine wave to the output voltage waveform. Each pulse in the PWM is then set in width, independent of each other, as necessary to satisfy the feedback loop, and maintain the output voltage waveform, as close as the gain of the loop will allow, to follow the reference waveform. This real-time type of tracking provides for an excellent output waveform, with comparatively little output filter, while providing both waveshape and voltage regulation at the same time. Another advantage to this type of control over the programmed control, as shown above, is the use of a full-sine rather than half-sine reference. With the half-sine it is necessary to add some additional control circuitry to steer the PWM to the proper drive circuit in order to generate each half of the output sine wave. By using a full-sine-wave reference, as depicted in Fig. 13-14, the additional circuitry is not necessary. The error amplifier output is all that is required to determine which drive circuit is necessary to turn on, in order to make the desired output waveform.

There are many variations on all the themes presented here, depending on the actual topology of the power and drive circuit employed. Real-time comparison of the output voltage waveform to a sine-wave reference is the most effective approach to synthesizing a sine-wave output. But the topology of the power circuit can again make a big difference. For example, the circuit shown in Fig. 13-15 uses a high-frequency DC-to-DC converter to boost a low-voltage battery up to a level that can be directly switched, to generate a sine-wave output. The immediate advantage to this circuit is the elimination of any low-frequency magnetics in the output. By generating a plus and minus DC bus that is directly modulated to for the sine-wave output, there is no further need for isolation, or stepup transformation, the two main purposes of the low-frequency transformer commonly found in the output of inverters. What this circuit has effectively done is replaced the low-frequency transformer with the entire boost converter. It is up to the designer to determine whether the one will cost-effectively replace the other, and at the power level being dealt with.

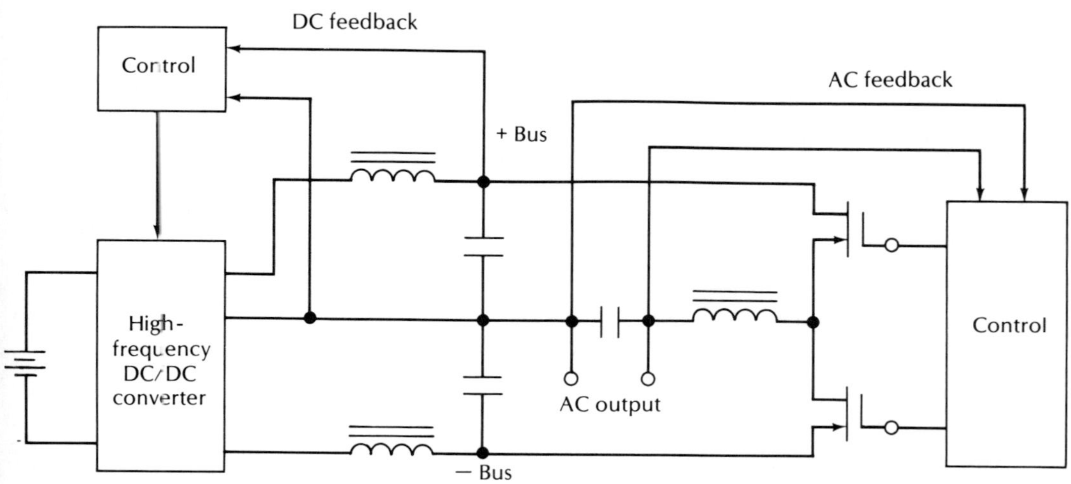

FIGURE 13-15 Boost converter provides high-voltage plus and minus bus to operate half-bridge inverter. No line frequency magnetics are required.

Two possibilities arise with the topology illustrated in Fig. 13-15. The first is the fairly straightforward approach of making a filtered DC bus, plus and minus, of approximately 170 V. Then applying any of a number of techniques for control and drive, the sine-wave output is generated. Real-time feedback can be employed as described above. This requires two feedback control circuits, though. The first is the control loop for the DC-to-DC converter itself, and the second is the feedback loop for the inverter. This can be reduced down to one loop, adding simplicity to the overall design. First, replace the real-time feedback control in the inverter with the programmed waveform technique described earlier. Adjust the sine reference and triangle waveforms as required to provide the desired output voltage, given some plus and minus voltage bus as provided by the DC-to-DC converter. Now, instead of regulating the DC-to-DC converter from its own DC output, bring the AC output around and use that to regulate the DC-to-DC converter. The benefit of having an AC voltage feedback here is a small feedback transformer can be utilized for isolation and scaling of the output voltage, without need of optocouplers. This method also provides a nice soft start feature for high inrush loads by allowing the inverter to begin delivering load voltage immediately as the DC-to-DC converter starts ramping up its output.

Another variation on this same theme is to make the inverter a square-wave drive, at the desired output frequency, rather than attempting to provide PWM control. Then use a sine reference in the DC-to-DC converter, rather than the conventional fixed DC voltage reference, and feed back the output AC as described above. This method will provide half-sine waveforms from the DC-to-DC converter, in both polarities. The square-wave drive to the inverter will then be synchronized with these half-sine waveforms to alternately pass every other one to the output load, effectively "unfolding" the waveform. An obvious advantage of this method is the elimination of complex control circuitry in the inverter portion of the converter, and the elimination of the output filter, no longer needed since the DC-to-DC converter is now generating the sine waveforms, and its own filter is all that is necessary. One problem with this method is sizing the filter capacitors to allow the waveshape to be maintained at light loads. Another problem is that now the DC-to-DC converter must be sized to handle the peak load current, compared with making the DC-to-DC converter output a straight DC output, where the filter capacitors handled these currents. That will add cost to the converter, and may make it a less desirable choice.

CHAPTER 14

System Configuration

Products intended to provide blackout protection, utilizing batteries and some form of DC-to-AC power conversion, have been around for many years. Typically these products are referred to as *uninterruptible power systems* (UPS) or *standby power systems* (SPS). The distinctions between the two types, their advantages and their disadvantages are outlined below.

UPS

For purposes of discussion it is important the understand the distinction between what is referred to as an "SPS" and what is a "UPS." Figure 14-1 shows a functional block diagram of a UPS. The arrows depict the flow of current during normal operation, loss of AC input operation, and bypass operation. By definition an uninterruptible power system is one that will display no loss of power to the load when normal AC input power is lost. The output voltage is uninterruptible. As shown in Fig. 14-1*a*, the power line is used to operate a battery charger, which, in turn, provides enough current and voltage to keep the battery charged and run the DC-to-AC inverter at full load. The inverter in this case can be any one of the examples shown in Chapter 10 in the section on power converter circuits. The battery charger could utilize one of several technologies from thyristors operated by phase control to the most advanced offline switching power supply design. The technologies used, for purposes of discussion right now, are not important. During normal operation, then, power flow is from the power line, through the charger, through the inverter, and out to the load. When the power line fails, as depicted by the arrows in Fig. 14-1*b*, then the battery maintains the current flow into the inverter to keep the load operating, until power returns and the battery charger takes over again and recharges the battery. Figure 14-1*c* shows the flow of power during bypass operation. Most UPS products will have a bypass switch, either standard or as an option, to allow the load to operate off the power line directly for any one of several reasons. This topic will be discussed in greater detail later.

There are several advantages to having the inverter in a UPS operating full-time. It can provide a great deal of immunity from power-line noise and

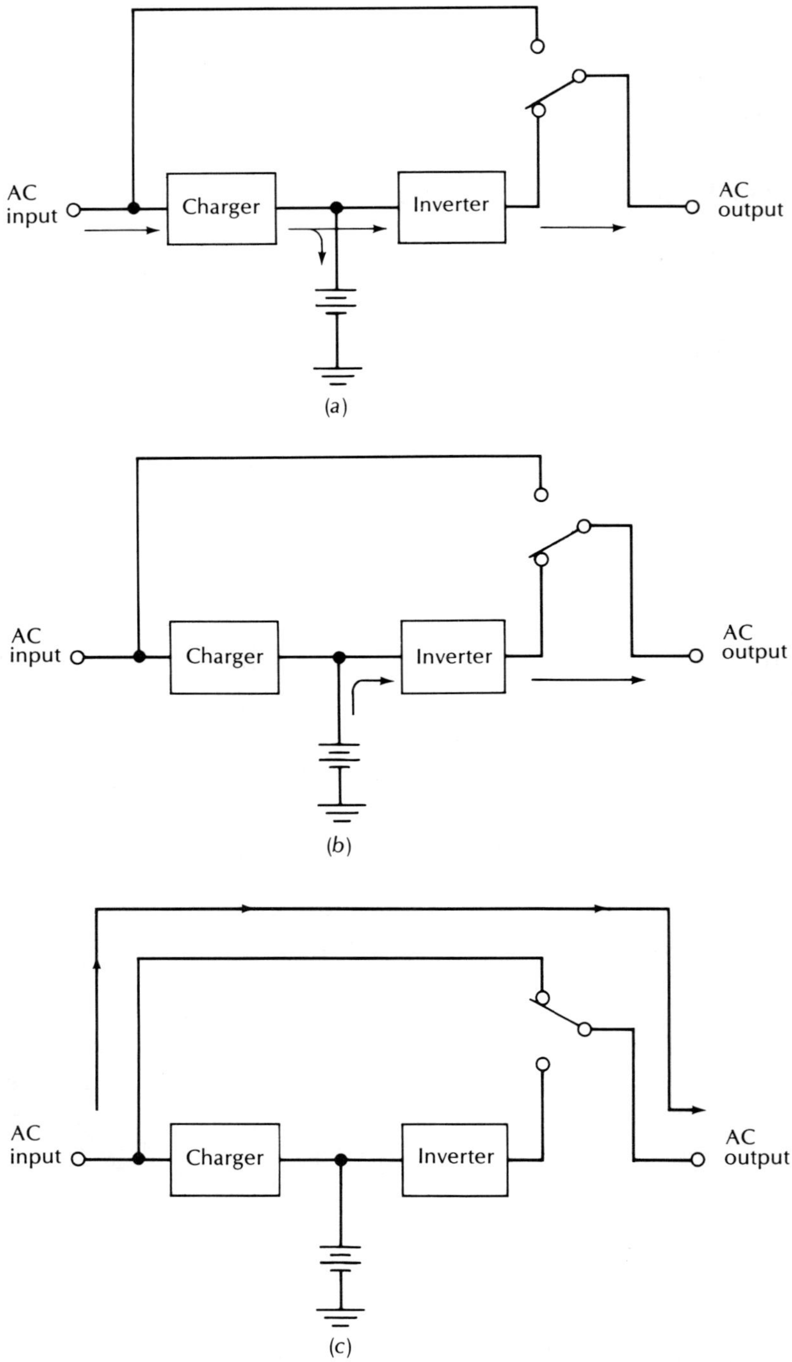

FIGURE 14-1 UPS block diagram depicting power flow during (*a*) normal operation; (*b*) loss of AC line; (*c*) bypass operation.

surges. A regulated battery charger that can withstand the stress of repeated surges as defined in IEEE-587 will provide an excellent buffer between the power line and the inverter. Assuming that the input to the inverter is clean, there is little more to do to ensure well-filtered and well-regulated AC output to the load. Having a well-regulated output voltage is another advantage of this type of system. Some loads are sensitive to sudden changes in power-line voltage, or may not operate over a very wide input voltage range. So in addition to providing line regulation for these loads, a UPS will provide continuous backup and immunity from surges. Another benefit in some UPS products is regulation of the frequency to the load. Some loads are frequency sensitive and will not operate properly when the input frequency fluctuates. Normally this is not too much of a concern when operating on the power grid within the continental United States. But if the power source is a generator, then the frequency may vary quite a bit. Most UPS products operate with the AC output phase-locked to the input line frequency. This will help ensure a frequency stability, during normal operation, that is at least as good as the power line. More sophisticated systems will provide a frequency detection circuit that will allow the phase-lock loop to shift the output frequency at only a limited rate of change, while tracking the input, and then desynchronize completely if the frequency moves outside preset limits. These types of detection circuits are usually only found on larger UPS products. Smaller units usually provide only the option of synchronization to the power line, or not.

It is seldom found that a product designed to operate from a 115-V-AC input was also designed in such a way that it makes tight input regulation of voltage and frequency mandatory. The input power supply in most products will usually tolerate very wide input operating conditions while providing tightly regulated output voltage to run the rest of the product. It is a fact of life that the power line is going to fluctuate, and in doing so will have an effect on a products operation, unless the designer takes some precautions against it. The precaution is rarely to specify a line regulator or UPS be placed ahead of the product. So, it is safe to say that most well designed products that are intended to operate from the power line will do quite well under most conditions. That fact pretty well eliminates the need for many of the benefits derived from a UPS. The conventional UPS design depicted in Fig. 14-1 suffers from a couple of disadvantages. First, it takes two power converters, operating full time, to provide full protection. There may be problems with audible noise associated with this full-time operation, unless the converters operate above 20 kHz. Continuous operation at high frequencies gives rise to RFI, both conducted and radiated, that must be dealt with, just as designers of switching power supplies must do. Continuous operation also implies design considerations must be made for cooling the system and providing a product with a high mean time between failure (MTBF). It is most desirable, when using a UPS ahead of a critical load, that the UPS not be the major cause of down time and lost data.

The losses in the inverter, plus the charge current that must be provided to the battery, may cause the input converter, or battery charger, to be rated as much as 20–40% greater in power than the inverter. The losses in the input converter, then, added to the rest, can make the AC-to-AC conversion efficiency quite poor. If both converters ran 80% efficiency, the total system would only be 64%. At high power levels the expense of operating the UPS, and throwing away that much power, is not desirable. At low power levels, where the cost of operating the UPS is not much more than the cost of operating the load alone, other considerations come into play. For safety, the most continuous current that should be taken from a 15-A wall outlet is 12 A. As the load current rating of a UPS approaches 12 A, the efficiency will be the limiting factor in getting safety agency approval, and still having the convenience of using a 15-A plug to connect to the AC line. It is desirable to be able to run any single load that could operate directly from a wall plug without having to hire an electrician to hard-wire the UPS to a higher current line. This limitation is not based only on efficiency, either. The technology used to effect the power conversion may also cause the RMS input line current to be greater than the power rating would imply. For example, a switching power supply operating off the power line, with large input filter capacitors and no form of power factor correction, may be rated at 200-W output. If it has an efficiency of 80%, then the input power would be 250 W. If the input current were sine-wave current, then $I = P/E$ would yield an input requirement of 2.17 A. But the input current will really have a large crest factor, which causes the RMS content of the waveform to go up. Instead of 2.17, the number could be 3.5 A RMS. In order to provide this high-crest-factor current to the load, the UPS must be rated higher than would be necessary if it were a resistive load. Now, if the input of the UPS is also a switching power supply, it will have the same effect on the power line as the load did. By the time all these currents add up, it can exceed the 12 A limit rather quickly, while the load by itself did not. That causes problems for most small UPS users, when they can not just set it in place and plug it in.

Overall, the concept of a UPS as a single solution to a lot of power-line problems is neither new, nor is it going to go away because of the problems and limitations pointed out here. There are some variations to the more conventional approach, as pointed out in chapter 10 discussing power converter circuits. Any circuit topology is going to have limitations or compromise for cost or other considerations. The final approach to the end product is not just one of meeting design goals but is also heavily dependent on establishing just what those goals must be. If it is determined that a UPS is the single best solution to meet the goals, and the associated problems are not limitations, then it is certainly a good solution. It has been for many years, and will continue to be.

SPS

A standby power system (SPS) differs from a UPS in one very important way. The power converter that is used to convert battery voltage into AC voltage in a typical SPS is sitting in an off state, waiting to turn on in the event of a power-line failure. Now, here is where it gets difficult. First, if a standby system, no matter how slow or fast it may switch on to replace a failed power line, will switch on fast enough to operate the load, without the load ever noticing that the power line failed, was the output uninterrupted and can it then be called a "UPS"? Second, if the standby system switches on fast enough that no noticeable load voltage waveform disturbance occurs, when the power line fails, is that considered uninterrupted and can it then be called a "UPS"?

The answer to both these questions is "maybe." There is really no standard developed that clearly defines what a UPS really is. There are standards used by safety agencies to differentiate standby from continuous operation. Specifically, if the unit operates with the main power converter in standby then it is technically a standby or SPS product. There are many manufacturers who call their standby product a UPS, mostly because the name implies more capability than an SPS typically will provide. For example, a ferroresonant line regulator that switches over to an inverter operating in standby mode is technically an SPS. But the slight ride-through capability of the ferro causes the output voltage to take on characteristics of a UPS. Figure 14-2 shows how a typical SPS works. The arrows show the power flow during normal operation in Fig. 14-2*a* and during a power-line failure in Fig. 14-2*b*.

There are several advantages of standby systems over the online type of system shown in Fig. 14-1. The battery charger in the standby system need not be any larger than is required to charge the battery, without running any current into the inverter. An inverter operated in a standby configuration, but designed to the same standards as that of an online UPS, will display a much greater MTBF since it turns on and supports the load only occasionally. Size and weight of the standby system can usually be reduced significantly over that of a UPS, as long as similar technologies are used.

Standby operation also has its own set of disadvantages. The most obvious one is the need to quickly detect a failing power line and get the converter started. This requires a very sensitive circuit. Care must be exercised in the design of the AC voltage detector circuit in order to prevent unwanted inverter operation when noise or load switching "glitches" appear on the power line. Unnecessary inverter operation will deplete the battery and it may not be fully charged when it is really needed. The lack of power line filtering or regulation has already been discussed, and whether it is really a problem is totally application-specific. Generally, it is not.

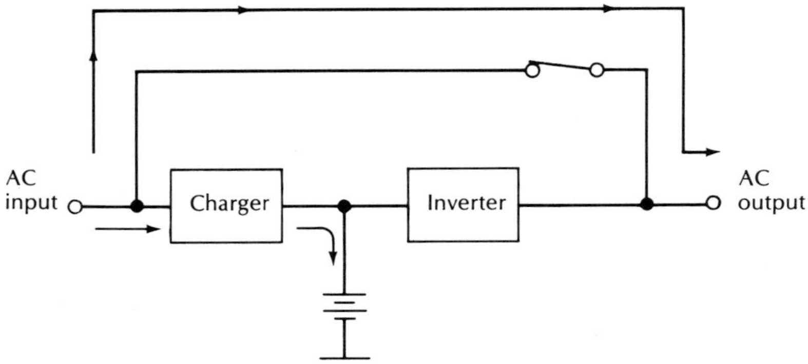

(a)

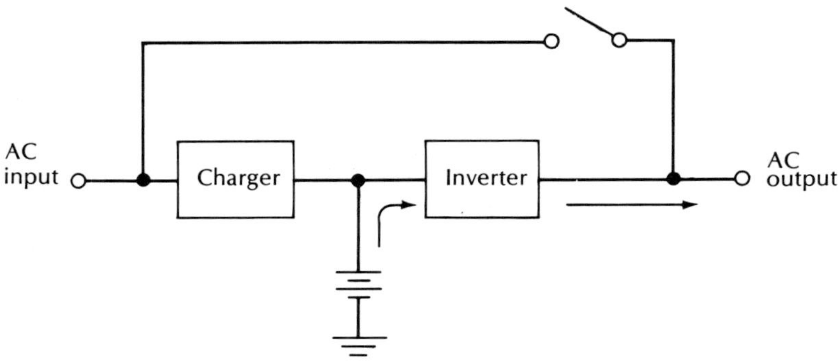

(b)

FIGURE 14-2 SPS block diagram depicting power flow during (*a*) normal operation; (*b*) loss of AC line.

Bypass Switch

Both circuits shown above utilize a switch mechanism to connect the load to the power line. In either case the switch can be a mechanical relay, or an electronic switch can be used. Electronic switches can be made up of any of the devices discussed in earlier chapters. Typically a thyristor, either a triac or inverse parallel SCRs, is used when the speed of an electronic switch is needed. Be careful applying an electronic switch to make effect transfers between the power line and the inverter. If the power line is present, then the inverter output should be phase-locked to the line before a transfer is made. This will help ensure that the load has minimal disturbance during the switching action. It is also important to be sure that the possibility of operating the inverter back into a dead power line does not exist. The result could be loss of inverter output if the dead power line looks like an overload.

Figure 14-3 shows an electronic switching circuit for a UPS system. There are switches in both the inverter output, which is normally on, and in the bypass line, which is normally off. This type of switching, usually called a *static transfer switch*, performs the same function as a form-C relay, where the arm of the relay is connected to the load and the poles connect to the inverter output and the bypass line. The difficulty with a relay is the switching time it takes to cause the load to be transferred from one source to the other. The electronic switch can be configured to provide virtually instant switching. The major problem encountered with application of electronic switching, using thyristors, is that they don't turn off when gate signal is removed. Instead, they will continue to conduct until the current goes to near zero. When the load consists of reactive elements, the current can be leading or lagging the voltage waveform. Some designs look for a voltage zero crossing before switching the static transfer switch from one source to the other. If the current is still going through the first switch, then it will cause problems when the second switch is turned on, unless precautions are observed. Referring to Fig. 14-3, assume that the inverter is phase-locked to the power line. When the load is switched between the two sources, current to or from the load, during each half-cycle, will be supplied by the SCR pointed in the same direction. For example, positive half-cycle current will flow through the SCR pointed toward the load, and negative half-cycle current will flow through the SCR pointed away. Assume for the moment that the load is unity power factor, and right at the peak of the positive half-cycle the load is switched from the bypass line to the inverter. The SCRs drive circuit can drive both SCRs on at the same time, or independently. If they both turn on at the same time, then there is a problem. If the power-line voltage is higher than the inverter voltage, then current will flow back through the SCR pointed away from the load into the inverter, even though the other SCR is supposed to carry the load current. This is because, as mentioned earlier, the current continues to flow through the SCR in the bypass line, even though it is turned off, until current goes to zero. Given the same situation as above, but instead the SCRs are controlled independently, let's try it again. Now when the bypass switch is turned off and the inverter side is turned on, only the SCR that is pointed toward the load is gated on. If the power-line voltage is higher, then the inverter will not supply any load current. If the inverter voltage is higher, then

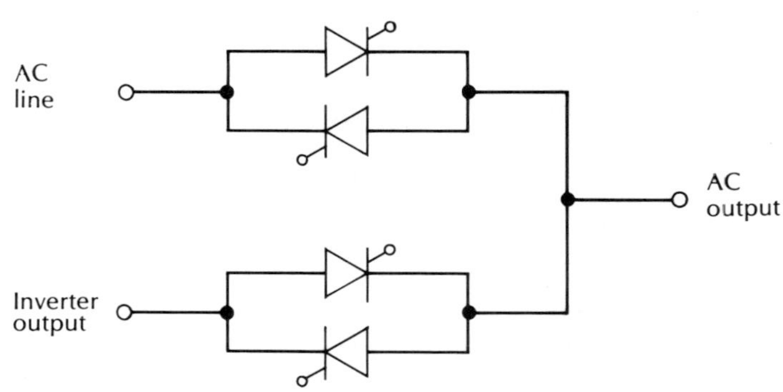

FIGURE 14-3 Inverse parallel SCRs arranged to form a static transfer switch.

the inverter will instantly take over the load and the SCR in the bypass line will be commutated off.

That solves the problem of preventing the wrong SCR from turning on, but there is still the problem of how to deal with displacement power factor load current. The solution is really quite simple. Instead of monitoring voltage zero crossings to determine when to switch the SCRs on, monitor the load current for both magnitude and direction. Use that information to determine which SCR must be on, and when, in order to switch smoothly. This method has been proved to work quite well and will prevent problems while allowing instant switching between sources. This same approach will work equally well with both an SPS or a UPS design. The main difference is with an SPS only half of the static switch is needed since the inverter is off, and its output can be connected directly to the load. Then, when the power line fails, the inverter can be started as soon as the load current goes to zero, indicating the static switch is now blocking and it is safe to run the inverter. Some consideration should be given in the design for how to handle a no-load situation, where there is no current to monitor. There are also safety agency considerations to be made. It is likely that any agency, whether U.S. or international, will require some form of mechanical break in the bypass line, particularly where the UPS or SPS is designed with a plug in power cord. If someone could remove the cord from the wall outlet, and then touch the end and get a shock from the operating inverter voltage because the static transfer switch failed and short-circuited, that is not acceptable. Furthermore, even during normal operation that same shock hazard can appear without the failure of any component. Most switching power supplies employ an input line filter that has capacitors from each line to ground. If the neutral line is not opened when the bypass switch opens, and the inverter is running, then the capacitors in the filter will act as a voltage divider and half of the inverter voltage will appear between the neutral line and the green wire ground. That voltage will exceed the limits allowed for voltage and current by safety agencies. While it is not likely to be lethal, it could impose serious problems and must be considered as part of the overall design of the switching mechanism.

Battery

Modern power protection systems, whether UPS or SPS, utilize sealed, maintenance free batteries, almost exclusively, to power levels in excess of 5 kW. At higher power levels it is common to find more conventional wet-cell batteries in use. Often at really high power levels the battery will take up a full room, and certainly more floor space than the UPS system it supports. But the size aspect does not change much at low power levels. The space taken up by the internal battery in most UPS products today will usually amount to as much or more space than is taken up by the electronics. The weight of the battery will usually be greater than the combined weight of all the electronics as well. Along with this size and weight goes cost. Battery cost will be one of the single most expensive items making of the material cost of a UPS.

The great size, weight, and cost of batteries makes it extremely important to properly select and implement a battery that will serve the requirement of the UPS system and provide a long useful service life. In a properly designed UPS system, the battery will be one of the weakest links just because of is susceptibility to misuse in charging, discharging, storage at elevated temperatures, and handling. Because they are sealed, maintenance-free batteries are even more at risk since there is no way to perform periodic maintenance. For this reason the following considerations are provided in the selection, charge, and discharge of sealed batteries. Most manufacturers of these batteries provide characteristic data that indicates approximately the same operating performance for similar ratings. The following is generalized for typical characteristics, and specific characteristics should be obtained from the manufacturer before making final design decisions.

Battery Characteristics

Sealed batteries are, as their description implies, sealed against spilling or loss of electrolyte, when operated within specification. The construction will allow operation in any position. Generation of gas within the battery is controlled to allow recombination of over 99% of the gas generated during normal use. These batteries are equipped with a low pressure venting system that will release excess gas and reseal automatically in the event that gas pressure rises to a level above the normal rate. While the sealed battery is typically considered safe to operate within enclosed areas, the low-pressure venting capability will still allow some gas to escape under certain conditions. Therefore, it is important to observe all the same safety considerations that must be observed when normal wet-cell batteries are used, particularly during charging. These batteries are particularly suited to UPS service, where deep discharge and cyclic use are common, because of the use of heavy lead calcium-alloy grids.

Typically the manufacturers of sealed batteries will specify a life of 3–5 years in float service, and depending on the depth of discharge, well over 1000 discharge and recharge cycles can be expected. High temperature is a real killer of batteries. When a system is designed for internal batteries, it is best to keep as much space between the batteries and sources of great heat, such as transformers and power transistor heat sinks. While it is not often practical, or even possible to do, the battery life will be maximized if it is kept in an operating temperature range of between 20 and 25°C. This is sometimes a little easier to do in an SPS, where the inside ambient temperature of the enclosure may be only 7 or 8°C above the outside temperature and increase only for short times when the inverter is called into service. A UPS, on the other hand, is operating continuously, and provision must be made to protect

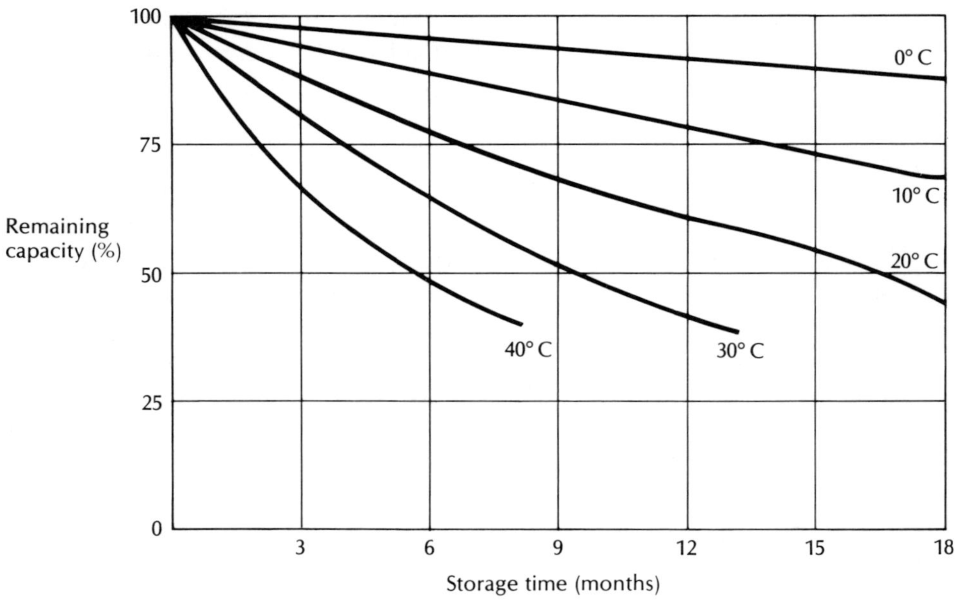

FIGURE 14-4 Self-discharge characteristics at various storage temperatures.

the battery from excessive heat buildup. Placing the battery near an air inlet, for example, rather than near an exhaust vent will improve the situation because cooler room air will be pulled over the battery as it enters the electronics enclosure.

The self-discharge of these batteries is also affected by the surrounding temperature. Figure 14-4 illustrates the remaining capacity versus storage time at various temperatures ranging from 0 to 40°C. As seen from these curves, after only 3 months at 40°C the battery is reduced to less than 75% of its full capacity, while at 0°C the battery has nearly full capacity. Lead acid batteries all display one undesirable characteristic. When allowed to sit for long periods of time while discharged, a lead sulfate will form on the negative plates. This "sulfation" will act as an insulator against the ability of the battery to accept a charge. The worse the sulfation gets, the less charge the battery will accept. When the battery is in service and is discharged, it is usual for the battery charger to recharge the battery within a short time of its discharge, and thus the phenomenon of sulfation never occurs. But, as seen from Fig. 14-4, even a fully charged battery that is in storage will self-discharge, and quite fast at higher temperatures. So, even during storage the effect of self-discharge can introduce sulfation and reduce the capacity of the battery before it is ever placed in service. It is highly recommended that products that utilize lead acid batteries, and are sold through distribution where it may be several months before the end user places the product in service, have the capability and instructions for periodic recharge to prevent this problem. This problem is also likely to affect batteries in inventory waiting to be used. If the inventory storage area is not refrigerated, it is best to keep the batteries as low to the floor as possible, since the warehouse will likely be hotter near the ceiling. Placing them directly on a concrete floor is a big help. Using a first-in–first-out rotation will also help.

Temperature during battery charge is also an important consideration. When the temperature rises, electrochemical activity within the battery increases. As the temperature falls, the activity decreases. In order to ensure maximum service life of the battery, it is best to use a temperature compensated battery charger that will automatically reduce charging as the temperature increases, and increase charge as the temperature decreases. Typical compensation is on the order of ±4 mV/°C per cell, with the standard center point is around 20°C. In actual numbers what this means is that a 12-V battery that would be normally charged at 2.25 Volts per cell or 13.5 V, at 20°C, should instead be charged at only 2.23 Volts per cell or 13.38 V, when the temperature increases by 5°C. While this is an ideal way to manage the charge voltage on the battery in varying environments, in reality few manufacturers of UPS products actually apply any means to temperature-compensate battery charge. Typically, if the product is intended for use in a protected environment, such as an office, the temperature within that space will be held pretty constant. Knowing the temperature rise within the electronics enclosure of the UPS will generally by a sufficient guide for selecting an adequate charge voltage.

Battery Selection

Above all else, the main consideration for selecting a battery to power a UPS is the minimum required run time necessary to operate the product at full load. Consider for a moment that the battery, while necessary to operate the product during power failures, just sits there the rest of the time. Power failures that draw on the battery are infrequent in most parts of the country, and in most applications the battery is never fully discharged before the power either returns or the user shuts down the load and the UPS in anticipation of the power failure lasting longer than the batteries are intended to support. So, here we are with a product containing a component that is extremely critical to the operation of the product, but only once in a while, and usually for short periods. But this component, just by its nature, is large, heavy, and quite expensive. Just for these reasons it is best to minimize the standard operating run time the product will provide and, if necessary, provide an interface to allow the connection of additional external batteries that can be added on if the occasion arises. This will minimize the burden of your standard product.

Most UPS manufacturers will provide enough battery to operate at full load for periods from 6 to 12 min. The wide range is due in part to the availability of batteries in limited capacity ratings versus the rating of the inverter at full load. An internal battery that provides a full-load run time of 10 min in a

500-W inverter may be the same choice, for cost reasons, in a 650-W inverter. But now the run time provided at the higher power level is less. A typical selection of 12-V batteries from one manufacturer includes three ratings at 6.0, 15, and 24 ampere-hour (AH) capacity. An inverter design determined to provide 7 min of run time with the 6.0-AH battery will discover that the additional cost and size implications of going to the next-higher capacity are not desirable. Another possibility is to series up two 6-V batteries that provide 10-AH capacity each, resulting in a 12-V, 10-AH battery, but that, too, adds cost and weight that may be beyond what is acceptable.

When selecting a battery for UPS application it is first necessary to consider the application of the UPS. If it is known that the product will be in an environment that will require the battery to support the load quite often, and for fairly long periods at a time, it is best to specify a battery that is rated for much more capacity than otherwise needed. This type of operation is considered to by "cyclic," and under these conditions the service life of the battery can be dramatically shortened. Consider that a battery has reached its end of useful life once its capacity has deteriorated to 60% of its AH rating. In cyclic use a battery that is only discharged 30% depth of discharge can expect to live for greater than 1000 discharge–recharge cycles. At 50% depth of discharge that number is down to less than 500, and at 100% depth of discharge less than 200 discharge–recharge cycles can be expected before the battery life is used up. If each of these cycles occurs only once per week, then the battery may well last 3–4 years, or longer. But at once every day or two it will be lucky to last a year if deeply discharged each time. It is wise to then specify a battery that will have to discharge only 30–50% in order to meet the required run time.

Of course, the load and operating conditions have a great effect on the life of the battery, and in commercial equipment it is seldom known just what the actual conditions will be. In this case it is necessary to make a decision regarding cost and performance and select the battery accordingly. Typical discharge characteristic curves for sealed lead acid batteries are shown in Fig. 14-5. These curves provide the expected run time from a battery, at 20°C, when discharged at a specified current and to a specified voltage. Each curve has a number followed by a *C*. This is the capacity figure; for example, 1*C* would be a discharge current of 10 A from a 10-AH battery, 2*C* would be 20 A, and so on. As the curves indicate, batteries that are discharged at higher rates can be discharged to a lower end voltage without damage, while at lighter discharge rates the end of discharge voltage increases. It is seen that the curves are nearly vertical at the end of discharge, so very little additional run time can be achieved from exceeding the curve. But severe damage can result from repeated operation that exceeds the specification.

The curves in Fig. 14-5 are useful in a couple of ways. First, once the actual discharge current of the inverter at full load is known, the curves will dictate what the discharge characteristic looks like for a particular battery capacity.

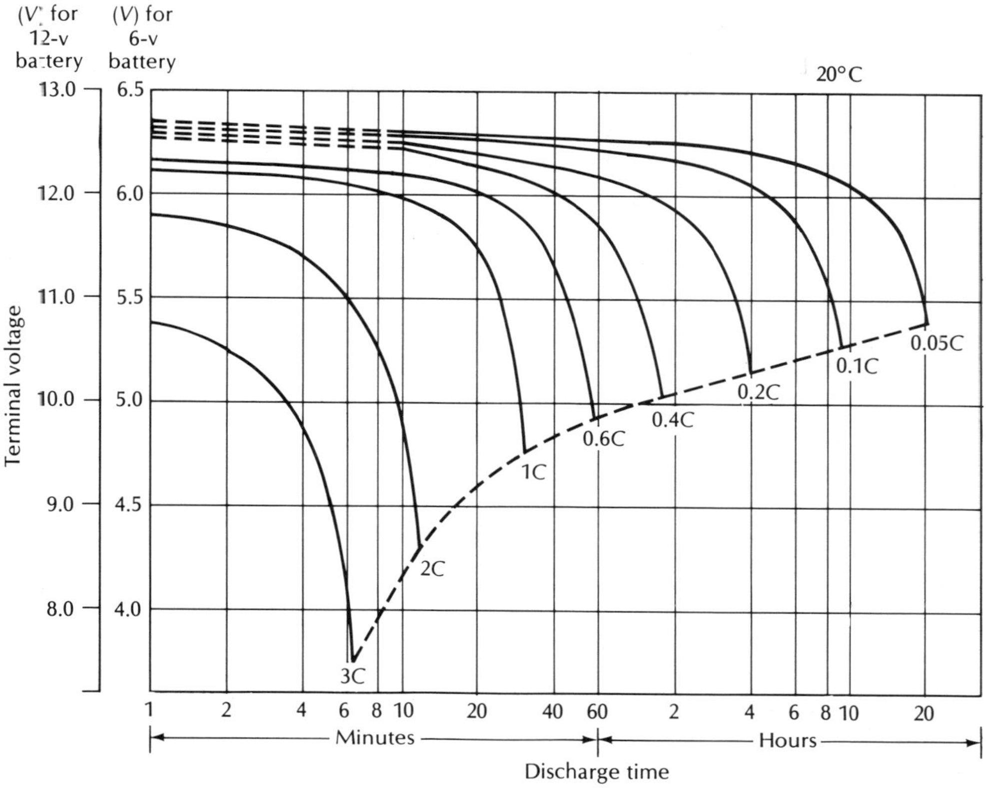

FIGURE 14-5 Typical discharge characteristic curves.

For example, a battery is required to provide 20 A for a minimum of 20 min. Locate the 20-min line along the bottom of the chart and follow it up to find that a 1*C* rating will be more than adequate. Either two 10-AH batteries could be placed in parallel, or a single 24-AH battery will work. The second way of using these curves is to determine what performance a specified battery will give. Say that you have a 24-AH battery in a UPS and you need to know how it will perform at half load. First determine what the discharge current will be. In this case lets use 4.8 A or 0.2*C*. From the curve it can be seen that about 4 of run time can be expected. In another example a 10 AH battery is being discharged at 30 A. The 3*C* curve indicates about 6 min of run time to 100% discharge. Operating these batteries beyond the 3*C* rate is not recommended, nor, as can be seen from the curves in Fig. 14-5, do you get much for your money. The internal impedance of these batteries pretty well limits their effectiveness to discharge rates of 3*C* or less. For applications requiring very high discharge rates and very short time runs, say, 2 or 3 min or less, it is best to consider some other battery technology, such as nickel–cadmium. Batteries of this type are better suited for very high discharge rates and will provide better service life under those conditions.

Battery Charging

Type of Charging

Battery charging falls into three basic categories: (1) constant-voltage charging, (2) constant-current charging, and (3) taper charging. Because of the severe nature of taper charging and the resulting damage that can result from its use, it will not be considered here. The service life of sealed lead acid batteries will be severely reduced if the proper precautions are not observed when charging the battery. One precaution is temperature and its effect on the charge voltage requirement. This was discussed above in detail under "Battery Characteristics." Aside from the voltage and current considerations it is important to make sure that the proper type of charger is used for the application.

Constant-Voltage Charging

Constant-voltage charging is the most common method used for charging this type of battery. In standby (float) use the charge voltage should be set to 2.25–2.30 V per cell. Cyclic use requires a slightly higher setting of 2.40–2.50 V per cell, although it is best to cut back the charge to the float setting after 24 h. Figure 14-6 shows an example of a constant-voltage charger.

Constant-Current Charging

Constant-current charging will allow a large amount of current to flow during the initial stage of charging, and then decrease as the charging progresses. When charging at 2.30 V per cell, charging current at the final stage of charg-

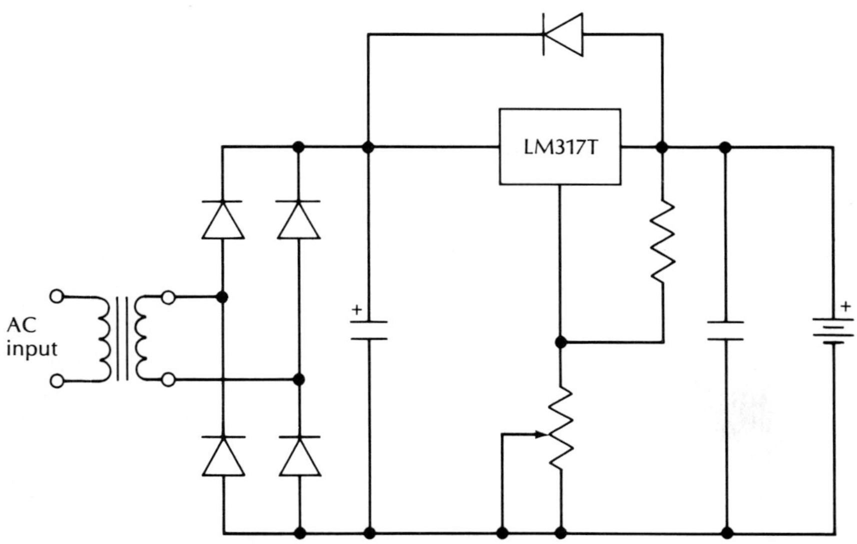

FIGURE 14-6 Simple adjustable battery charger utilizing a common three-terminal adjustable regulator.

ing will drop to as little as 0.002–0.005CA. Under this type of charge when a battery is charged up to the level of 100% of the discharged ampere-hours, the electrical energy stored and available for discharge will be 90% or more of the energy applied during charging. These batteries will accept a high-charging current at the initial stage of charging. Although it is recommended that the charging current be normally limited to 0.25CA, internal construction of the battery will not allow the battery to accept more than about 2CA, and the charging current will be reduced to a relatively small value in a very brief period of time. Usually, in standby use, no current limit is needed because of this self-limiting feature.

The most desirable charging method is a two-step charger. This type will initially charge by constant current. The charge voltage rises until it reaches a preset 2.45 V per cell, at which time the charging mode automatically changes to a constant voltage. Once the charge current reduces to a sufficient level, the voltage automatically reduces to 2.30 V per cell, and float charges the battery. This method will provide the best charge recovery in the shortest time while not stressing the battery unduly.

Battery Maintenance

All batteries will lose capacity through self-discharge while sitting unused. They should periodically be topped off in order to maintain their useful life. Generally within 6 months of manufacture, a 4–6-h constant current of 0.1CA, or 15–20 h at constant voltage of 2.40 V per cell, will suffice. Between 6 and 12 months, increase the times to 8–10 h at constant current, or 20–24 h at constant voltage. Before attempting to top off a battery that has been out of use for more than 12 months, check its open-circuit voltage. If it is not at least 2.0 V per cell, it is probably not worth the effort.

Batteries that have been overdischarged may have some difficulty in recharging. The initial indication of charge current may be quite low, giving cause for concern that the battery is bad, or possible fooling a two-step charger into reducing the voltage to the lower setting and not putting a proper charge into the battery. The internal resistance of an overdischarged battery will cause the initial stage of charge current to be quite low, but it will rise rapidly over a period of approximately 30 min. After that normal charging characteristics should be observed, if the battery wasn't damaged from misuse.

Overall, the battery selected to operate a UPS is extremely important. It is just as important to select a properly sized battery from a reliable manufacturer as all the rest of the electronics that goes into the design. The entire single point of purchasing a product to provide backup power, whether a SPS or UPS, is that it must provide that power when the power line fails. Without the battery to make all the electronics work, all that is left is a lot of electronics. Sealed batteries can leak, can go bad sitting at a distributor, and can spoil from contamination introduced at the time of manufacture. Unfortunately, there is little that can be done to determine that the battery installed in a UPS will provide reliable power for a reasonable period of time. If it provides the predicted run time at full load at the time of installation, then at least it is good at that time. If care in charging and storage is maintained, then all that is left is to trust the manufacturer. It is difficult to do life tests on batteries to determine which manufacturer to go with. That could take years. Often the best solution is to see what the high-volume users of these types of batteries are using, and go with that same battery. They may have done your life testing for you.

Index